Enzymes in Neurodegenerative Disorders

Bijo Mathew • Della Grace Thomas Parambi
Editors

Enzymes in Neurodegenerative Disorders

Mechanism and Therapeutic Potentials

Springer

Editors
Bijo Mathew
Department of Pharmaceutical Chemistry
Amrita School of Pharmacy, Amrita Vishwa Vidyapeetham
Kochi, Kerala, India

Della Grace Thomas Parambi
Department of Pharmaceutical Chemistry
Jouf University
Sakaka, Saudi Arabia

ISBN 978-981-97-6821-9 ISBN 978-981-97-6822-6 (eBook)
https://doi.org/10.1007/978-981-97-6822-6

This Springer imprint is published by the registered company Springer Nature Singapore Pte Ltd.
The registered company address is: 152 Beach Road, #21-01/04 Gateway East, Singapore 189721, Singapore

Preface

We are happy and proud to bring you the second edition of our neurology book *Enzymes in Neurodegenerative Disorders*. According to the book's authors, Bijo Mathew and Della Grace Thomas Parambi, there has been a tremendous increase in knowledge over the last few decades regarding the basic biological causes of neurological disorders, which are always connected to enzymes in the central nervous system. The etiology of many neurological diseases is multifactorial, leading to disruptions in neuronal circuits that can distort cognitive, neurobiological, and psychological systems. Enzymes, particularly modulatory enzymes, have always been an area of thrust in treating neurodegenerative diseases and have been exploited as a therapeutic option for treating the brain diseases. Enzymes can also be opted as a therapeutic regimen in a multidimensional way for treating various neurodiseases.

A neurologist/neurochemist knows the advances and ongoing discoveries in modern neurological practices, but this does not provide the basis for excellence in clinical practice. Finding the sweet spot between role of enzyme in neurological disease and practice is the goal of this book. This project begins with a firm grounding in the principles of anatomy, physiology, and biochemistry, focusing on the enzymes involved in the neurological pathways that are essential for understanding neurological symptoms and signs that are artfully extracted, processed, and abstracted by the various academicians in neuroresearch.

This book mainly highlighted in general neurodegenerative diseases and enzymes in general, the role of enzymes in major neurological disorders like Alzheimer's disease, Parkinson's disease, ALS, epilepsy, hunting's disease, prion disease, stroke and also future directions and challenges in enzyme research where enzyme potential can be used as a drug target or biomarker. An abnormal enzyme is most often a direct result of an underlying genetic abnormality. They can also be considered as potential biomarkers. We can also learn from identifying defective enzymes, as it provides new molecular targets to focus on for developing more effective pharmacotherapeutic options. Identifying and screening for these genetic abnormalities can be used for early identification and prevention of disease in individuals with a genetic predisposition. Throughout its length history, multiple incarnations of enzyme therapy have been tried. The modern advances in genetic engineering show significant promise for correcting these abnormalities and developing revolutionary cures, despite ongoing ethical concerns.

Given the crucial role enzymes play in the neuropathology of various neurological disorders, understanding their mechanism is essential for developing better treatment modalities and screening techniques.

Kochi, Kerala, India — Bijo Mathew
Sakaka, Saudi Arabia — Della Grace Thomas Parambi

Contents

1 **Introduction to Enzymes in Neurodegeneration** 1
Chameli Ratan, Akash Marathakam, and M. K. Unnikrishnan Mazhuvanchery

2 **Role of Enzymes in Alzheimer's Disease** 21
Sumitra Patel, Shweta Shrivastava, Parameswary V. Marar, Rajan Swami, and Manish Kumar Jeengar

3 **Role of Enzymes in Amyotrophic Lateral Sclerosis** 57
S. Maya and M. Shamnas

4 **Enzymes Involved in Multiple Sclerosis** 77
Jobin Jose and Priyam Purkait

5 **Role of Enzymes in Epilepsy** 89
Manal Mohammed, P. Prabitha, and Tariq G. Alsahli

6 **Role of Enzymes in Huntington's Disease** 105
Malavila B. Nair, Gouri Shyna Rajesh, Della Grace Thomas Parambi, Arafa Musa, Tariq Ghazi Alsahli, Siju Ellickal Narayanan, Sandeep Bindra, and Bijo Mathew

7 **Role of Enzymes in Parkinson's Disease** 113
Aathira Sujathan Nair, Sonia Dangi, Arun K. Thomas, and Mahendran Sekar

8 **Role of Enzymes in Prion Disease: A Molecular Mechanism** 127
Sayed Mohammed Firdous, Sourav Pal, and Subhajit Mandal

9 **Modulation of Several Downstream Cascades Served by Enzymes in the Pathogenesis of Stroke** 153
Sayed Mohammed Firdous and Sourav Pal

10 **Future Directions and Challenges in Enzyme Research for Neurodegenerative Diseases** 171
Karishma Manohar Rathi, Priyanka Vilas Pingat, and Sayed Mohammed Firdous

Editors and Contributors

About the Editors

Bijo Mathew is a Professor in the Department of Pharmaceutical Chemistry at the Amrita School of Pharmacy, Amrita Vishwa Vidyapeetham, Kochi, India. Dr. Bijo Mathew obtained his M.Pharm in Pharmaceutical Chemistry from The Tamil Nadu Dr. M.G.R. Medical University, Tamil Nadu, India, in 2008, and Ph.D. in Pharmaceutical Sciences from Jawaharlal Nehru Technological University, Hyderabad, India, in 2016. Mathew's primary focus areas are the drug discovery of neurodegenerative disorders, cancer, and molecular modeling. His main research interest lies in the design of enzyme inhibitors, especially MAOs, ChEs, and BACE1. He was recognized as one of the Top 2% Scientists globally by Stanford's list from 2021 to 2023. He has published 230 papers (h-index = 43, 6011 citations on Google Scholar; h-index = 40, 5070 citations on Scopus) as of July 2024. With 16 years of teaching and research experience, Mathew has authored one book, edited one book, contributed to seven book chapters, and delivered 34 invited talks at national and international levels. Dr. B. Mathew serves as a chief executive guest editor of the journals *Molecules*, *Current Topics in Medicinal Chemistry*, and *Current Drug Targets*. He also serves on the editorial boards of journals *Brain Sciences and Combinatorial Chemistry* and *High Throughput Screening*. For the development of MAO inhibitor drugs, Dr. Mathew secured funding from several government funding organizations in India, including the DRDO and ICSSR. He received the Amrita Vishwa Vidyapeetham Chancellor's Publication and Merit awards in 2021 and 2024. He has received three patents from the Korean and German Intellectual Property offices.

Della Grace Thomas Parambi is currently working as an Assistant Professor in the Department of Pharmacy, Jouf University. She has completed her Ph.D. in Pharmaceutical Analysis. She previously served as the Head and Principal of Triveni Institute of Pharmacy, Thrissur. She has more than 20 years of teaching experience in analytical chemistry, medicinal chemistry, molecular modeling, and drug design. Her primary research activities are focused on designing MAO/ChE inhibitors for neurodegenerative disorders. Several of her publications have received the "Most Downloaded Paper Award" for the years 2018–2020. She has also published more than 70 research articles in peer-reviewed international journals and authored one book and six book chapters. She is a member of many international scientific societies and organizations, including the Indian Society of Science and Technology. Additionally, she serves as an EVC expert in the All India Council for Technical Education (AICTE). She has received numerous awards, including the AIAF-IEDRA Indian Leadership Award for Education Excellence for the year 2014 and also the IAF Best Performing Institution of the Year Award in 2014.

Contributors

Tariq Ghazi Alsahli Department of Pharmacology, College of Pharmacy, Jouf University, Sakaka, Saudi Arabia

Sandeep Bindra Department of Pharmaceutical Chemistry, Amrita School of Pharmacy, Amrita Vishwa Vidyapeetham, AIMS Health Sciences Campus, Kochi, India

Sonia Dangi Jeffrey Cheah School of Medicine and Health Sciences, Monash University Malaysia, Subang Jaya, Selangor, Malaysia

Sayed Mohammed Firdous Department of Pharmacology, Calcutta Institute of Pharmaceutical Technology & AHS, Howrah, West Bengal, India

Manish Kumar Jeengar Department of Pharmacology, Amrita School of Pharmacy, Amrita Vishwa Vidyapeetham, AIMS Health Sciences Campus, Kochi, Kerala, India

Jobin Jose Department of Pharmaceutics, NITTE Deemed-to-be University, NGSM Institute of Pharmaceutical Sciences, Mangalore, Karnataka, India

Subhajit Mandal Department of Pharmaceutical Sciences, Faculty of Science and Engineering, Dibrugarh University, Dibrugarh, India

Parameswary V. Marar Department of Pharmacology, Amrita School of Pharmacy, Amrita Vishwa Vidyapeetham, AIMS Health Sciences Campus, Kochi, Kerala, India

Akash Marathakam Department of Pharmaceutical Chemistry, Sree Gokulam SNGM College of Pharmacy, Thuravoor, India

Bijo Mathew Department of Pharmaceutical Chemistry, Amrita School of Pharmacy, Amrita Vishwa Vidyapeetham, Kochi, Kerala, India

S. Maya Department of Pharmacology, KMCT College of Pharmacy, Malappuram, Kerala, India

Manal Mohammed Department of Pharmaceutical Chemistry, KMCT College of Pharmacy, Mampara, Malappuram, Kerala, India

Arafa Musa Department of Pharmacognosy, College of Pharmacy, Jouf University, Sakaka, Saudi Arabia

Aathira Sujathan Nair School of Pharmacy, Monash University Malaysia, Subang Jaya, Selangor, Malaysia

Malavila B. Nair Department of Pharmaceutical Chemistry, Amrita School of Pharmacy, Amrita Vishwa Vidyapeetham, AIMS Health Sciences Campus, Kochi, India

Siju Ellickal Narayanan Department of Pharmacology, College of Pharmaceutical Sciences, Government Medical College, Kannur, India

Sourav Pal Department of Pharmacology, Seacom Pharmacy College, Jaladhulagori, Sankrail, West Bengal, India

Della Grace Thomas Parambi Department of Pharmaceutical Chemistry, Jouf University, Sakaka, Saudi Arabia

Sumitra Patel Department of Pharmacology, Amrita School of Pharmacy, Amrita Vishwa Vidyapeetham, AIMS Health Sciences Campus, Kochi, Kerala, India

Priyanka Pingat Department of Pharmacy Practice, Dr. D.Y. Patil Institute of Pharmaceutical Sciences and Research, Pimpri-Chinchwad, India

P. Prabitha Department of Pharmaceutical Chemistry, JSS College of Pharmacy, JSS Academy of Higher Education & Research, Mysore, Karnataka, India

Priyam Purkait Department of Pharmaceutics, NITTE Deemed-to-be University, NGSM Institute of Pharmaceutical Sciences, Mangalore, Karnataka, India

Gouri Shyna Rajesh Department of Pharmaceutical Chemistry, Amrita School of Pharmacy, Amrita Vishwa Vidyapeetham, AIMS Health Sciences Campus, Kochi, India

Chameli Ratan Department of Pharmacy Practice, Amrita School of Pharmacy, Amrita Vishwa Vidyapeetham, AIMS Health Science Campus, Kochi, Kerala, India

Karishma Rathi Department of Pharmacy Practice, Dr. D.Y. Patil Institute of Pharmaceutical Sciences and Research, Pimpri-Chinchwad, India

Mahendran Sekar School of Pharmacy, Monash University Malaysia, Subang Jaya, Selangor, Malaysia

M. Shamnas Department of Pharmacy Practice, KMCT College of Pharmacy, Malappuram, Kerala, India

Shweta Shrivastava School of Pharmacy, School of Health and Allied Sciences, ARKA JAIN University, Jamshedpur, Jharkhand, India

Rajan Swami Chitkara College of Pharmacy, Chitkara University, Rajpura, Punjab, India

Arun K. Thomas Department of Pharmacology, Annoor Dental College, Muvattupuzha, Kerala, India

M. K. Unnikrishnan Mazhuvanchery Department of Pharmacy Practice, Amrita School of Pharmacy, Amrita Vishwa Vidyapeetham, AIMS Health Science Campus, Kochi, Kerala, India

Introduction to Enzymes in Neurodegeneration

1

Chameli Ratan, Akash Marathakam, and M. K. Unnikrishnan Mazhuvanchery

Abstract

Neurodegenerative diseases are a group of disorders characterized by progressive degeneration of the nervous system, leading to a decline in cognitive function, movement control, or both. This chapter elucidates the essential role of enzymes in the development of neurodegenerative diseases (NDDs), particularly Alzheimer's disease (AD), focusing on enzymes such as neprilysin (NEP), endothelin-converting enzyme (ECE), insulin-degrading enzyme (IDE), and plasmin, emphasizing their roles in amyloid-beta (Aβ) metabolism and therapeutic strategy development. Reduced levels of NEP and ECE in AD patients lead to pathological Aβ accumulation, suggesting that upregulating these enzymes could alleviate AD symptoms by promoting Aβ clearance. IDE, which degrades both insulin and Aβ, is examined in the context of its role in insulin resistance and AD pathology, suggesting therapeutic benefits from upregulating IDE activity. Plasmin, a protease involved in Aβ fibril degradation, shows reduced activity in AD, highlighting the need for interventions to enhance its function. This chapter also discusses the impact of polyglutamylating enzymes on microtubule dynamics and neuronal stability, identifying their dysfunction as a contributor to neuronal damage, and advocating a potential therapeutic target. Furthermore, the renin-angiotensin system (RAS), specifically the enzymes angiotensin-converting enzyme (ACE) and ACE2, is discussed in relation to neuroinflammation and NDDs, with evidence suggesting that ACE inhibitors and angiotensin receptor blockers might provide neuroprotective effects. The importance of lifestyle fac-

C. Ratan · M. K. Unnikrishnan Mazhuvanchery (✉)
Department of Pharmacy Practice, Amrita School of Pharmacy, Amrita Vishwa Vidyapeetham, AIMS Health Science Campus, Kochi, Kerala, India

A. Marathakam
Department of Pharmaceutical Chemistry, Sree Gokulam SNGM College of Pharmacy, Thuravoor, India

B. Mathew, D. G. T. Parambi (eds.), *Enzymes in Neurodegenerative Disorders*,
https://doi.org/10.1007/978-981-97-6822-6_1

tors such as diet and physical activity in managing NDDs is also emphasized. This comprehensive examination of enzymatic roles and lifestyle impacts lays the groundwork for developing multifaceted strategies to combat neurodegeneration.

Keywords

Neurodegenerative diseases · Enzymes · Amyloid-beta · Alzheimer's disease · Protein aggregation

1.1 Introduction

Neurodegenerative disorders (NDDs) are clinical presentations defined by gradual, irreversible dysfunction and degeneration of neurons in the central and peripheral nervous systems (Jellinger 2009; Gan et al. 2018). Aging and age-related complications are the main risk factors for NDDs (Guerreiro et al. 2020). The complex and multifactorial degenerative processes include oxidative stress (OS), excitotoxicity, mitochondrial dysfunctions, and autophagy impairment (Gan et al. 2018). The pathogenic mechanisms involved in NDDs are OS leading to the formation of free radicals and reactive oxygen species (ROS), fragmentation of the neuronal Golgi apparatus, abnormal protein dynamics (misfolding, aggregation, defective degradation, and proteasomal dysfunction), impaired bioenergetics, dysfunction of neurotrophins, various neuroinflammatory and neuroimmune processes, mitochondrial dysfunctions, DNA damage, and disruption of cellular/axonal transport (**see** Fig. 1.1).

The above pathogenic processes are interconnected in complex cycles, forming the molecular cascades causing NDD and ultimately resulting in cellular

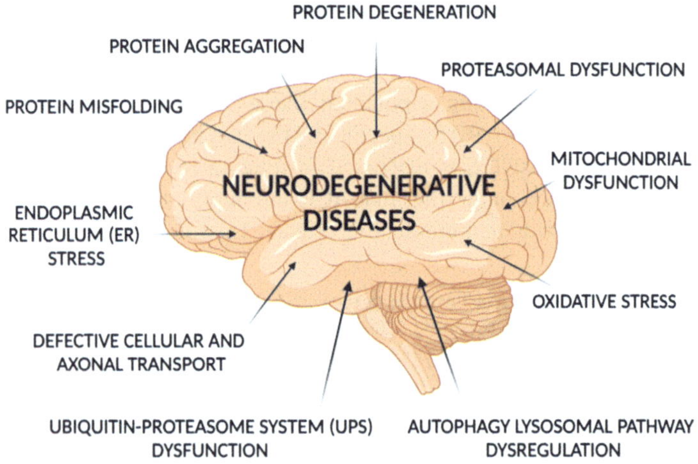

Fig. 1.1 Various mechanisms behind NDDs

dysfunction and death (Jellinger 2010). Aging reduces the levels of nicotinamide adenine dinucleotide (NAD), an anti-inflammatory and neuroprotective molecule present in the human brain and cerebrospinal fluid (CSF). Aging and the associated cellular senescence from the accumulation of senescent cells in the brain increase the CD38 enzyme levels, leading to neuroinflammation and NDDs (Guerreiro et al. 2020). Current treatments focus on symptom management rather than addressing the underlying pathological changes, leaving a significant clinical gap. A notable approach involves computer-aided drug design, showing promise to create innovative therapeutics that specifically target the molecular pathways implicated in NDDs (Salman et al. 2021).

1.2 Pathogenic Mechanisms of Neurodegenerative Diseases

Protein aggregation in NDDs arises from anomalous protein interactions, resulting in the accumulation of misfolded proteins that self-aggregate. The specific protein involved in the affected neurons establishes distinctive clinical presentations characteristic of the disease (Jellinger 2010). Abnormal protein–protein interactions initiate vicious circles that contribute to the dysfunction and death of glial and neuronal cells, which in turn causes the disintegration of neuronal networks (Palop et al. 2006; Seeley et al. 2009). Protein aggregation mainly occurs from the conversion of highly soluble brain proteins into insoluble polymers (Ecroyd and Carver 2008; Morris et al. 2009). The "Toxic Oligomer" hypothesis demonstrates the causative connection between the formation of protein aggregates and NDD. According to this proposition, NDDs are believed to be directly associated with the harmful effects of compounds generated in early stages, notably soluble oligomers and protofibrillar variants originating from misfolded proteins (Jellinger 2010). Protein misfolding in NDDs involves conformational changes due to genetic defects, amino acid substitutions, or repeats (Schwartz and Ciechanover 2009) (**see** Fig. 1.2). NDDs are associated with the accumulation of aberrant proteins generated by abnormal synthesis, impaired degradation, OS, and altered post-translational modifications (Jellinger 2010).

The protein folding and formation are governed by a comprehensive system known as the proteostasis network (Balch et al. 2008). The preservation of proteostasis relies on a system of pathways that regulate various aspects such as protein folding, synthesis, aggregation, trafficking, disaggregation, and degradation (Hutt et al. 2009). Certain genetic disorders characterized by amino acid substitutions involve loss-of-function mechanisms, leading to the elimination of aberrant proteins (Winklhofer et al. 2008). However, when misfolded proteins are not eliminated, they accumulate to form toxic oligomers contributing to gain-of-function pathogenesis, as observed in Huntington's disease (HD), Alzheimer's disease (AD), and Parkinson's disease (PD) (**see** Fig. 1.3) (Mirzaei and Regnier 2008).

Endoplasmic reticulum (ER) stress is implicated in NDDs, such as AD and PD (Jellinger 2010). ER stress induces distress in both the structure and function of the ER, resulting in the buildup of misfolded proteins and disturbances in calcium

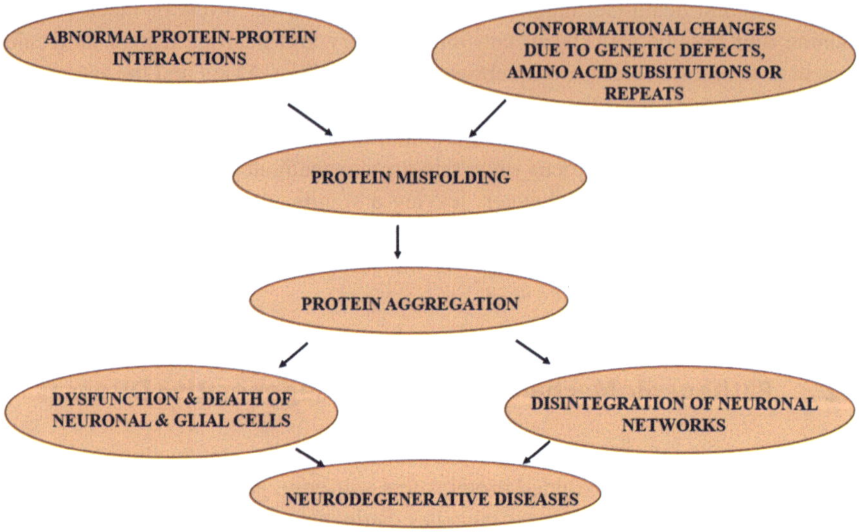

Fig. 1.2 The flow chart that explains how protein misfolding and protein aggregation lead to NDD

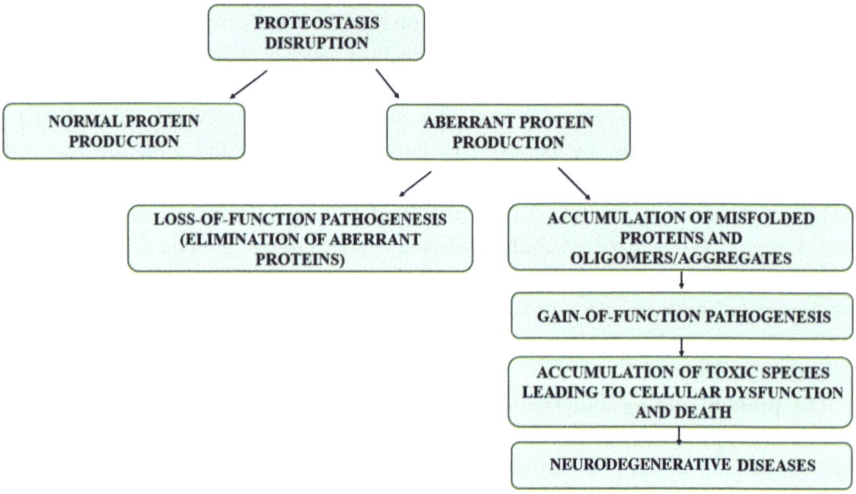

Fig. 1.3 The diagram shows the role of proteostasis disruption in NDD

balance, together contributing to synaptic dysfunction, impaired plasticity, and neurodegeneration (Bezprozvanny 2009) (**see** Fig. 1.4). To avoid this, the unfolded protein response will be triggered to alleviate ER stress and promote survival through pathways involving endoplasmic reticulum kinase, inositol-requiring kinase-1, and activation of transcription factor 6 (Schroder 2008).

The autophagy-lysosome pathway (ALP) and the ubiquitin-proteasome system (UPS) regulate the protein turnover (Lehman 2009). The UPS, a substrate-specific

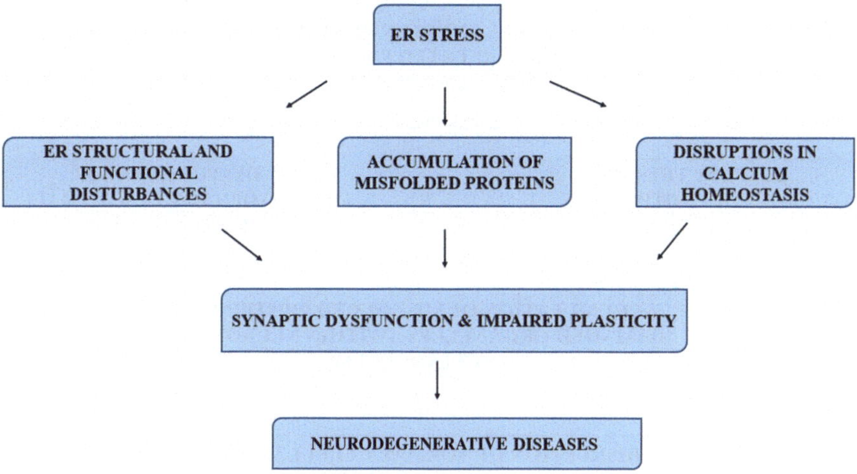

Fig. 1.4 The relationship between ER stress and NDD

proteolytic system, functions as an intracellular mechanism for clearing proteins (Jellinger 2010). The main feature of the UPS is the ubiquitination process with the help of 26S proteasome and Ub-activating enzymes (Schulman and Harper 2009), leading to the regulation of various downstream signaling pathways (Ye and Rape 2009). Dysfunctional UPS may lead to the accumulation of misfolded proteins within cells, disrupting normal cellular activity, and this dysfunction can both cause NDDs or result from them (Jellinger 2010). ALP is a pathway important for protein clearance and organelle degradation (Martinet et al. 2009). ALP and UPS signaling pathways together protect against neurodegeneration (Tai and Schuman 2008). Dysregulation of macro-autophagy is associated with synaptic dysfunction and neuronal cell death (Chu et al. 2009) (**see** Fig. 1.5).

OS is another mechanism involved in NDD. OS occurs when the body's production of free radicals surpasses its antioxidant defense mechanisms, resulting in damage to biological molecules (Sultana and Butterfield 2009). ROS, including hydroxyl radicals and peroxynitrite, are potent agents of cellular damage, impacting proteins, nucleic acids, fatty acids, and lipids (Nunomura et al. 2009). Consequently, OS significantly contributes to the cellular pathology observed in NDDs (**see** Fig. 1.6). The accumulation of misfolded proteins exacerbates oxidative damage by disrupting the function of the proteasome (Halliwell 2006). Disruption in metal homeostasis, particularly the accumulation of iron in brain regions of NDD patients, catalyzes the generation of highly reactive radicals, further inducing oxidative damage (Smith et al. 2010). Post-mortem tissues from patients with neurodegenerative conditions such as amyotrophic lateral sclerosis (ALS), AD, and PD, among others, exhibit oxidative damage across lipids, proteins, and DNA (Migliore and Coppede 2009).

Mitochondrial dysfunction is a central theme in neurodegeneration because the mitochondria are both a source and a target of ROS (Dubinsky 2009). Mitochondrial damage affects respiratory chain enzymes and mitochondrial DNA, leading to

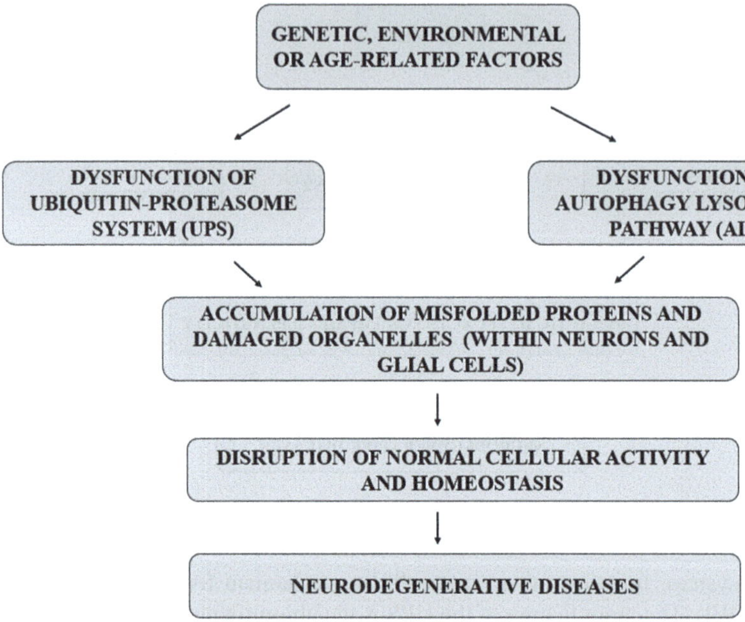

Fig. 1.5 The flow chart depicts the role of UPS and ALP pathways in NDD

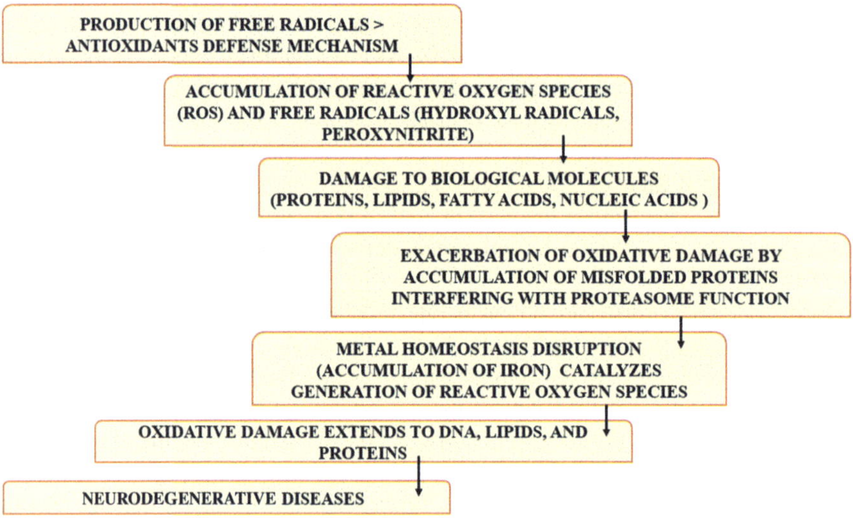

Fig. 1.6 The flow chart shows the relationship between oxidative stress and NDD

aberrant energy metabolism and thereby contributing to neuronal dysfunction (Yang et al. 2008). Mitochondrial dysfunction is a common feature in NDDs such as AD and PD (Swerdlow et al. 2014). In AD, oxidative damage and dysfunctional energy

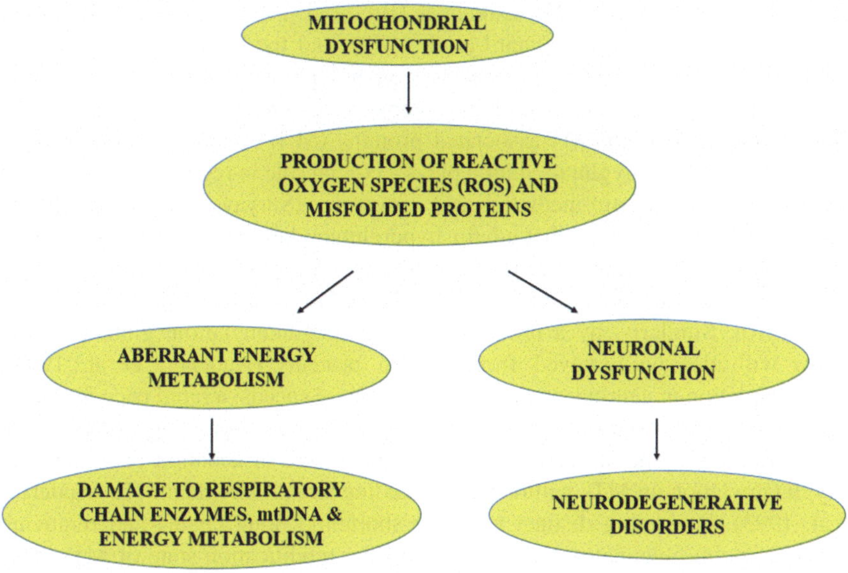

Fig. 1.7 The flow chart explains the role of mitochondrial dysfunction in NDD

metabolism contribute to a 50% reduction in mitochondrial RNA content, leading to decreased oxidative phosphorylation (Swerdlow and Khan 2009; Mancuso et al. 2009). The "mitochondrial cascade hypothesis" proposes that dysfunction in mitochondria serves as an initial catalyst in the pathophysiology of AD. Furthermore, oxidative harm to mitochondrial DNA (mtDNA) in AD might lead to irregular mitochondrial dynamics, dysfunction, and neuronal damage (Swerdlow et al. 2014) (**see** Fig. 1.7). In AD as well as PD, mitochondrial dysfunction gives rise to the generation of ROS, resulting in the impairment and misfolding of proteins (Davison et al. 2009). In AD, amyloid-beta (Aβ) plays a role in mitochondrial dysfunction (Hauptmann et al. 2009), while in PD, mutations in genes like Parkin (Davison et al. 2009) and PINK1 contribute to mitochondrial abnormalities (Yang and Lu 2009). Defective cellular and axonal transport are the other common mechanisms involved in NDDs. It results in disturbances within the microtubule (MT) system, motor proteins, and cargo transportation, ultimately resulting in gradual neuronal degeneration and diminished connectivity (Jellinger 2010).

1.3 Enzymes Involved in NDDs

1.3.1 Polyglutamylating Enzymes

Polyglutamylation, a posttranslational alteration, involves the addition of glutamate side chains, of varying lengths, to the γ-carboxyl groups of glutamic acid residues (Ruse et al. 2022). It is notably elevated in stable MT structures, such as those in neurons (Audebert et al. 1993), axonemes (Bré et al. 1994), centrioles (Bobinnec et al. 1998), basal bodies (Regnard et al. 1999), and the highly dynamic mitotic spindle (Regnard et al. 1999). Tubulins undergo polyglutamylation at their carboxy-terminal tails, which are exposed on the outer surface of assembled MTs, serving as binding sites for various MT-associated proteins (MAPs) and molecular motors (Kubo et al. 2010). Polyglutamylation plays a crucial role in regulating MT–dynein interactions and spastin-mediated MT severing (Suryavanshi et al. 2010). Consequently, it is proposed that tubulin hyperglutamylation might impact the interactions of MAPs and molecular motors with the MT cytoskeleton, ultimately contributing to neurodegeneration (Kubo et al. 2010; Suryavanshi et al. 2010; Lacroix et al. 2010). Similarly, by a process known as deglutamylation, glutamate side chains will also be removed from tubulin proteins (Audebert et al. 1993). Deglutamylation in neurons operates at two separate speeds: a rapid pace for shortening lengthy glutamate side chains and a slower pace for eliminating glutamates at branching points. This process intricately controls the distribution of mono- and polyglutamylation on MTs within cells, including along individual MTs (Audebert et al. 1993). By having distinct rates for shortening long chains and removing branching point glutamates, the cells control the modification state of MTs. This regulation plays an important role in various cellular processes, influencing MT dynamics, stability, and interactions with associated proteins (Audebert et al. 1993). The ability to selectively modulate the extent of polyglutamylation allows for fine-tuning the functional properties of MTs in different cellular contexts (Audebert et al. 1993). The enzymes responsible for polyglutamylation, known as polyglutamylases, belong to the tubulin tyrosine ligase-like (TTLL) protein family. Consequently, it is vital for cellular function to uphold appropriate levels of tubulin polyglutamylation through the coordinated activity of polyglutamylases and deglutamylating enzymes (Ruse et al. 2022; Rogowski et al. 2010) (**see** Fig. 1.8). Cytosolic carboxypeptidase 1 (CCP1), alternatively identified as Nna1, functions as a tubulin deglutamylase enzyme accountable for eliminating the C-terminal tyrosine from alpha-tubulin, leading to the loss of CCP1 enzyme function (Pang et al. 2023). Loss of CCP1 function has been linked to various defects in Purkinje cell degeneration (PCD) mice, including the degeneration of Purkinje cells, cerebellar granule neurons, thalamic neurons (Mullen et al. 1976), mitral cells in the olfactory bulb, and retinal photoreceptors, along with fertility issues (Wang and Morgan 2007). CCP1 plays an important role in modifying detyrosinated alpha-tubulin by removing its C-terminal glutamate residue. CCP1 will only cleave peptide bonds between two glutamate residues at the C terminal and not catalyze the initial detyrosination process of alpha-tubulin (Rogowski et al. 2010). The enzyme also regulates MT dynamics and function within the cell (Pang et al. 2023). CCP1 directly processes two substrates: myosin light-chain kinase 1 (MLCK1) and telokin. MLCK1 (130 kDa) is notably present in the uterus and telokin (20 kDa) in the stomach (Herring et al. 2006). These substrates support the proteolytic processing of gene-encoded acidic C termini (Tanco et al. 2015). Furthermore, the enzyme is also involved in the removal of posttranslational tubulin polyglutamylation leading to tubulin modifications, which affects the dynamic nature of MTs, and influences

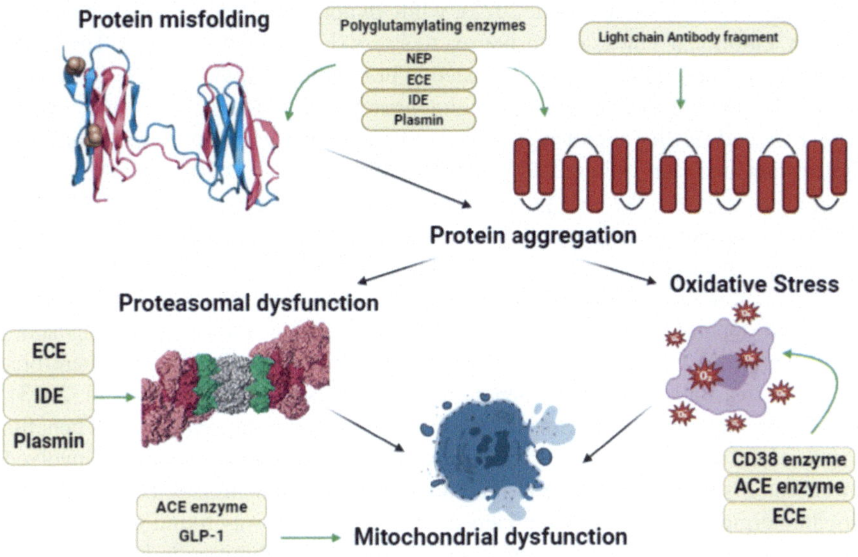

Fig. 1.8 The diagram represents the role of enzymes in NDDs

essential cellular functions such as cell division, intracellular transport, and the structural integrity of the cytoskeleton (Rogowski et al. 2010). CCP1 has been shown to increase levels of tubulin polyglutamylation in specific brain regions (like the olfactory bulb and cerebellum) undergoing neurodegeneration in PCD mice. This observation suggests that elevated tubulin polyglutamylation could be a contributing factor to neurodegeneration in PCD mice (Rogowski et al. 2010). Studies downregulating tubulin-specific neuronal polyglutamylases (TTLL1) in the cerebellum of young PCD mice demonstrated reduced Purkinje cell death and enhanced motor coordination. These findings establish that tubulin hyperglutamylation is a causative factor in Purkinje cell degeneration (Rogowski et al. 2010).

1.3.2 Angiotensin-Converting Enzymes (ACEs)

Renin is an enzyme primarily released by the kidneys. Its primary role is to process angiotensinogen, a protein synthesized in the liver into angiotensin peptides. The ACE and angiotensin-converting enzyme 2 (ACE2) are key enzymes within the renin-angiotensin system (RAS). They are responsible for the conversion of angiotensin I (Ang I) to angiotensin II (Ang II) and the subsequent conversion of Ang II to Angiotensin 1–7, respectively. Angiotensin II acts on two types of receptors: the AT1 and AT2 receptors. The RAS is responsible for regulating blood pressure, cardiovascular functions, and fluid balance in the human body (Fountain et al. 2023). ACE and ACE2 enzymes influence the pathophysiology of various NDDs (**see** Fig. 1.8). These enzymes cleave substrates like neurotensin and dynorphin. ACE cleavage of neurotensin effects dopaminergic transmission (Torruella-Suarez and

McElligott 2020; Tschumi et al. 2022), while dynorphin cleavage affects pain perception, emotional states, and responses to stress (Limoges et al. 2022). The RAS, particularly ACE and ACE2, is implicated in brain inflammation. ACE is associated with inflammation through ATR1 (angiotensin receptor I) signaling, while ACE2, Ang 1–7, and AT2R (angiotensin receptor II) form the protective axis, exhibiting antioxidant and anti-inflammatory effects (Miners et al. 2020). The ACE gene exhibits an insertion/deletion polymorphism (I/D polymorphism) that affects ACE mRNA expression and oligopeptidase activity (Santiago et al. 2023). The presence or absence of an Alu repeat sequence leads to insertion (I) or deletion (D) phenotypes, affecting ACE levels. The I/D polymorphism is associated with levels of serotonin and dopamine metabolites in the CSF. The D/D genotype individuals show higher serum ACE activity, potentially leading to lower levels of Ang II. Therefore, there exists a link between the genotype and ACE activity (Santiago et al. 2023). ACE inhibitors and angiotensin receptor blockers (ARBs) showed potential benefits in counteracting neurodegenerative diseases by the degradation of peptides involved in neurotransmission and by the modulation of neuroinflammation implicated in dopaminergic transmission (Santiago et al. 2023). ACE was found to be overexpressed in AD-affected brains, and administering ACE inhibitors reduced memory deficits, ROS formation, and Aβ plaque formation (Savaskan et al. 2001; Thomas et al. 2021; Torika et al. 2016). ACE inhibitors also improve cognitive functions by reducing neuroinflammation (Thomas et al. 2021; Torika et al. 2016). Dopaminergic neurons in PD are influenced by angiotensin II. It influences dopamine synthesis and release, leading to glial cell activation and upregulated RAS activity. Therefore, ATR1 antagonists showed neuroprotective action in the PD mouse model (Garrido-Gil et al. 2012). Additionally, ARBs also decreased dopaminergic cell loss in PD mouse models (Dauer and Przedborski 2003).

1.3.3 CD38 Enzyme

CD38 is found throughout the human brain, with notably higher levels in specific areas such as the cingulate anterior, putamen, caudate, thalamus, pallidum, and olfactory bulb (Quintana et al. 2019). At the cellular level, CD38 is present in microglial cells, astrocytes, and neurons (Guerreiro et al. 2020). CD38 enzyme regulates NAD levels in the brain. Studies conducted on CD38 knockout mice exhibited a tenfold increase in NAD levels, suggesting that CD38 is the key factor, which regulates intracellular NAD levels (Aksoy et al. 2006). The enzyme also regulates NAD bioavailability and affects the function of NAD-dependent enzymes like SIRT1 (Braidy et al. 2014). CD38 facilitates the transformation of NAD+ into cyclic adenosine diphosphate-ribose (cADPR), a widely used calcium mobilizer. Hence, the CD38 enzyme is important for regulating intracellular calcium levels (Higashida et al. 2007). The enzyme is also involved in the release of neurotransmitters, influencing oxytocin, vasopressin, norepinephrine, serotonin, and dopamine release (Martucci et al. 2019). Studies indicate that following neuroinflammatory insults (e.g., HIV-1 encephalitis), astrocytes increase the CD38 expression. Also,

CD38 knockdown mice generate less astrocyte proinflammatory cytokines and chemokines. These suggest that CD38 enzymes are involved in the regulation of neuroinflammatory processes (Kou et al. 2009) (**see** Fig. 1.8). NAD levels are consistently reduced in NDDs such as AD (Sonntag et al. 2017), PD (Wakade et al. 2014), ALS (Wang et al. 2017), and multiple sclerosis (Braidy et al. 2013). A decrease in NAD levels indicates the expression and activity of the CD38 enzyme (Camacho-Pereira et al. 2016). Such initial evidence indirectly suggests an association between increased CD38 expression and NDDs, making it a promising target for combating neuroinflammation and NDD (Lautrup et al. 2019).

1.3.4 Enzymes in AD

AD is a debilitating NDD characterized by neuronal loss, cognitive impairment, and eventual death (Lichtenthaler 2004). Deposition of Aβ was identified as the causative factor in AD (Lichtenthaler 2004; Chow et al. 2010). Aβ is a naturally occurring metabolite continuously generated and broken down in the brain (Lichtenthaler 2004). Aβ is produced through the cleavage of a membrane-bound protein called amyloid precursor protein (APP) by specific proteases, also termed secretases, which include α-secretase, β-secretase, and γ-secretase (Chow et al. 2010). The accumulation and aggregation of Aβ prompt an inflammatory response that leads to neuritic injury, tau protein hyperphosphorylation, and the development of fibrillary tangles, ultimately resulting in neuronal dysfunction and cell death (Koudinov 2004). If Aβ is cleaved by γ-secretase, then three Aβ peptide fragments, namely, Aβ40, Aβ42, and Aβ43, are formed, of which Aβ42 is more toxic and shows an increased risk of aggregation (El-Agnaf et al. 2000). Similarly, the main β-secretase that causes Aβ peptide formation is BACE1 (Cole and Vassar 2007). Evidence also indicates that Aβ may affect the function of voltage-dependent Ca^{2+} channels (Prista von Bonhorst et al. 2022) and stimulate GTPase activity in neurons, potentially causing long-term cognitive decline (Molnar et al. 2004). Aβ was also found to induce neurotoxic effects by activating caspases, effectors of apoptosis, and influenced synapses and synaptic structures, thereby affecting functional plasticity, which is crucial for memory and learning (Lautrup et al. 2019; Bredesen et al. 2010). The downregulation of certain enzymes known as amyloid-degrading enzymes was found to contribute toward the development of AD. The enzymes are discussed below:

1.3.4.1 Neprilysin (NEP)

NEP, a zinc metalloprotease initially identified as a kidney enzyme (Turner 1997), participates in the breakdown of various peptide hormones, adrenocorticotropin hormone, atrial natriuretic peptides, opioid peptides, bombesin-like peptides, tachykinins, and chemotactic peptides (Duvaux-Miret et al. 1992). NEP has been observed to degrade physiologically active peptides such as Aβ peptide (Iwata et al. 2000). In individuals with AD, both the mRNA and protein levels of NEP have been noted to decline (Yasojima et al. 2001), particularly in neurites surrounding β-amyloid plaques (Apelt 2003). Consequently, the decrease in NEP levels with aging may significantly contribute to AD development. An NEP homolog, NEP 2, has been identified, which exhibits Aβ-degrading capabilities and is more prevalent in regions relevant to AD pathology, hinting at a potential role in Aβ breakdown (Shirotani et al. 2001) (**see** Fig. 1.8). Thus, neprilysin emerges as a promising target for therapeutic development in AD treatments.

1.3.4.2 Endothelin-Converting Enzyme (ECE)

ECE is an analog of NEP. ECE-1, a variant of ECE, is a membrane-bound metalloprotease responsible for converting big endothelin into vasoactive endothelin (Barnes 1999). Studies suggest that ECE-1 degrades Aβ and regulates Aβ concentrations. In an experiment conducted on Chinese hamster ovarian cells, it was observed that overexpression of ECE-1 led to a significant reduction, up to 90%, in extracellular Aβ concentration. Additionally, mice with decreased ECE-1 activity displayed notably elevated levels of both Aβ40 and Aβ42 in the brain, indicating ECE's role in modulating Aβ concentrations (Eckman et al. 2001) (**see** Fig. 1.8). Therefore, enhancing ECE activity emerges as a potential therapeutic strategy for AD.

1.3.4.3 Insulin-Degrading Enzyme (IDE)

IDE, alternatively called insulysin, belongs to the zinc metallopeptidase group and is categorized within the inverzincin family of zinc proteinases (Leissring 2021). IDE serves as the primary enzyme responsible for degrading Aβ. Studies have demonstrated that the transgenic enhancement of IDE expression in neurons leads to a substantial decrease in brain Aβ levels and effectively prevents the formation of amyloid plaques (Leissring et al. 2003) (**see** Fig. 1.8). Similarly, genetic studies in mice showed that a fall in IDE levels also decreased Aβ degradation in the brain, thereby diminishing the cerebral accumulation of endogenous Aβ (Prince et al. 2003). IDE is also responsible for the clearance of intracellular Aβ (Schmitz et al. 2004), and hence, upregulation of the IDE protein levels or increasing its activity can be considered a novel therapeutic approach for the treatment of AD. Another function of IDE is breaking down insulin. In AD patients, Aβ decreases insulin binding to its receptor via direct competitive inhibition (Xie et al. 2002), thereby diminishing glucose metabolism (Ling et al. 2002) and defective signal transduction (Bigl et al. 2003). This causes insulin resistance (Ho et al. 2004) which further leads to upregulation of insulin receptors (Bigl et al. 2003) and downregulation of IDE and γ-secretase activity (Ho et al. 2004). Hence, insulin receptor signaling can modulate Aβ production, suggesting a potential mechanism for the observed association between insulin dysregulation and AD.

1.3.4.4 Plasmin

Plasmin, a thrombolytic enzyme found in the blood, originates from the inactive zymogen, plasminogen, under the influence of serine proteinases seered to as plasminogen activators. Its primary role involves breaking down components of the extracellular matrix and functioning as a thrombolytic enzyme within the bloodstream (Werb 1997). Decreased levels of plasmin have been observed in the brains of AD patients, particularly within the hippocampus (Ledesma et al. 2000). Purified plasmin effectively degrades both Aβ and Aβ fibrils by cleaving them at multiple sites. Studies have revealed that externally applied plasmin mitigates Aβ neurotoxicity (Tucker et al. 2000) (**see** Fig. 1.8). Plasmin has also been shown to interact with lipid rafts, which are cholesterol-enriched membrane microdomains (Ledesma et al. 2003). These microdomains facilitate the production of amyloid peptides (Ehehalt et al. 2003), indicating a potential connection between hyperlipidemia—a growing lifestyle concern—and the risk of developing AD.

1.3.4.5 Others

Fragments of light-chain antibodies exhibit proteolytic capabilities against Aβ (**see** Fig. 1.8). One of these fragments demonstrates an α-secretase-like function, generating specific Aβ fragments, while another exhibits carboxypeptidase-like activity, modifying Aβ aggregation and counteracting its cytotoxic effects (Rangan et al. 2003). These observations imply that the proteolytic breakdown of Aβ mediated by antibody fragments presents a potential therapeutic avenue for managing Aβ aggregation and toxicity in vivo (Liu et al. 2004). The reduction of endogenous Aβ levels has been associated with glucagon-like peptide-1 (GLP-1). Activation of neuronal GLP-1 receptors has been demonstrated to regulate neuronal plasticity and survival (Perry 2004). Both GLP-1 and its analog exendin-4 were found to diminish endogenous Aβ levels in mouse brains and reduce APP levels in the neurons (Perry 2004) (**see** Fig. 1.8). Consequently, this approach could be utilized for the development of a novel treatment strategy for AD.

1.4 Enzymes as Potential Therapeutic Targets in AD

Increasing the activity of Aβ-degrading enzyme, plasmin, NEP, EDE, and IDE, could be a promising therapeutic strategy for treating AD (Wang et al. 2010). Strategies that enhance α-secretase activity may reduce the production of Aβ peptides and mitigate the pathological processes associated with neurodegeneration (Sambamurti 2007). Inhibition of BACE1 is considered a therapeutic target for AD. Strategies for inhibiting human β-secretase activity include the development of peptidomimetic, non-cleavable substrate analogs. Antisense approaches using small inhibitory nucleic acids (siRNAs), such as antisense oligonucleotides, catalytic nucleic acids, and small interfering RNAs (siRNAs), are being explored for BACE1 downregulation (Nawrot 2004). Gene transfer via a recombinant adeno-associated viral vector demonstrated the axonal transport of NEP to presynaptic sites, resulting in reduced Aβ levels in mouse models (Iwata et al. 2004). Lentiviral vectors expressing human neprilysin also showed success in reducing Aβ deposits and ameliorating neurodegenerative alterations in transgenic mouse models of amyloidosis. While gene therapy offers a potential means to selectively elevate neuronal NEP levels, an alternative approach involves pharmacological upregulation of NEP (Marr et al.

2003). Green tea extract, for example, has been reported to have such an effect, and systematic screening for NEP regulators holds promise (Grimm et al. 2013).

1.4.1 Lifestyle Factors and NDDs: Understanding the Impact

Evolutionary characteristics of genes associated with AD were examined by analyzing the conservation of splice sites across different vertebrate genes. Surprisingly, nearly all genes associated with AD are evolutionarily old (Nitsche et al. 2015), suggesting a likelihood of a strong link between modern lifestyle and the surge in NDD. Protein-coding genes undergo a moderate rate of alterations, whereas non-coding genes experience more pronounced changes. The accelerated evolution of non-coding genes underscores their significance in disease processes. The heightened adaptability of genes associated with AD, particularly non-coding variants, suggests a rapid adjustment at a functional level. Recent evolutionary shifts in the human brain appear to correlate with fundamental aspects of neurodegeneration observed in AD, implying a potential connection between evolutionary dynamics in the human brain and the development of NDD (Nitsche et al. 2015). Hence, there is a need for a paradigm shift in how we conceptualize NDD-like AD. It is also important to consider the accelerated evolution and unique characteristics of AD-associated genes when developing strategies for disease modification. Modern lifestyle disorders such as insulin dysregulation and hyperlipidemia significantly heighten the risk of NDD-like AD due to their systemic effects on metabolism and vascular health. Insulin dysregulation, often occurring due to poor dietary habits and sedentary lifestyles, contributes to neuroinflammation and OS, accelerating cognitive decline. Concurrently, hyperlipidemia, exacerbated by high-fat diets and limited physical activity, disrupts brain lipid metabolism and undermines neuronal integrity, further predisposing individuals to AD and related NDDs. Addressing these lifestyle factors is crucial for mitigating the onset and progression of NDDs.

1.5 Conclusion

The process of neurodegeneration is complex, our understanding of enzymatic contributions provides a foundation for targeted interventions and emphasizes the importance of holistic approaches that address both molecular basis and lifestyle factors to alleviate the burden of NDD on individuals and society. The enzymes such as polyglutamylating enzymes, ACE enzymes, CD38 enzymes, NEP, ECE, IDE, plasmin, and other enzymes (light-chain antibody fragment and GLP-1) were found to have a significant effect in inducing NDDs. Therefore, targeted therapeutic interventions aimed at modulating these enzymatic pathways could help slow down the disease progression. However, the challenges lie not only in understanding the role of different enzymes in NDDs but also in addressing the broader context of modern lifestyle disorders. The surge in NDDs correlates with contemporary lifestyles characterized by poor dietary habits, sedentary behaviors, and environmental stressors, all of which exacerbate the risk of neurodegeneration. Hence, emphasizing lifestyle modifications, promoting neuroprotective behaviors, and advancing precision medicine initiatives tailored to individual enzymatic profiles may offer better possibilities in both preventive care and therapeutic interventions.

References

Aksoy P, White TA, Thompson M, Chini EN (2006) Regulation of intracellular levels of NAD: a novel role for CD38. Biochem Biophys Res Commun 345:1386–1392

Apelt J, Ach K, Schliebs R (2003) Ageing-related downregulation of neprilysin, a putative β-amyloid-degrading enzyme, in transgenic Tg2576 Alzheimer-like mouse brain is accompanied by an astroglial upregulation in the vicinity of β-amyloid plaques. Neurosci Lett 339:183–186

Audebert S, Desbruyeres E, Gruszczynski C, Koulakoff A, Gros F, Denoulet P, Eddé B (1993) Reversible polyglutamylation of alphaand beta-tubulin and microtubule dynamics in mouse brain neurons. Mol Biol Cell 4:615–626

Balch WE, Morimoto RI, Dillin A et al (2008) Adapting proteostasis for disease intervention. Science 319:916–919

Barnes K, Turner AJ (1999) The endothelin converting enzyme is located on αactin filaments in smooth muscle cells. Cardiovasc Res 42:814–882

Bezprozvanny I (2009) Calcium signaling and neurodegenerative diseases. Trends Mol Med 15:89–100

Bigl M et al (2003) Cortical glucose metabolism is altered in aged transgenic Tg2576 mice that demonstrate Alzheimer plaque pathology. J Neural Transm 110:77–94

Bobinnec Y, Moudjou M, Fouquet JP, Desbruyeres E, Eddé B, Bornens M (1998) Glutamylation of centriole and cytoplasmic tubulin in proliferating non-neuronal cells. Cell Motil Cytoskeleton 39:223–232

Braidy N, Lim CK, Grant R, Brew BJ, Guillemin GJ (2013) Serum nicotinamide adenine dinucleotide levels through disease course in multiple sclerosis. Brain Res 1537:267–272

Braidy N, Poljak A, Grant R, Jayasena T, Mansour H, Chan-Ling T, Guillemin GJ, Smythe G, Sachdev P (2014) Mapping NAD+ metabolism in the brain of ageing Wistar rats: potential targets for influencing brain senescence. Biogerontology 15:177–198

Bré MH, de Nechaud B, Wolff A, Fleury A (1994) Glutamylated tubulin probed in ciliates with the monoclonal antibody GT335. Cell Motil Cytoskeleton 27:337–349

Bredesen DE, John V, Galvan V (2010) Importance of the caspase cleavage site in amyloid-β protein precursor. J Alzheimers Dis 22(1):57–63. https://doi.org/10.3233/JAD-2010-100537

Camacho-Pereira J, Tarragó MG, Chini CCS, Nin V, Escande C, Warner GM, Puranik AS, Schoon RA, Reid JM, Galina A et al (2016) CD38 dictates age-related NAD decline and mitochondrial dysfunction through an SIRT3-dependent mechanism. Cell Metab 23:1127–1139

Chow VW, Mattson MP, Wong PC, Gleichmann M (2010) An overview of APP processing enzymes and products. Neuromolecular Med 12(1):1–12. https://doi.org/10.1007/s12017-009-8104-z

Chu CT, Plowey ED, Dagda RK et al (2009) Autophagy in neurite injury and neurodegeneration: in vitro and in vivo models. Methods Enzymol 453:217–249

Cole SL, Vassar R (2007) The Alzheimer's disease beta-secretase enzyme, BACE1. Mol Neurodegener 2:22. Published 2007 Nov 15. https://doi.org/10.1186/1750-1326-2-22

Dauer W, Przedborski S (2003) Parkinson's disease: mechanisms and models. Neuron 6:889–909

Davison EJ, Pennington K, Hung CC et al (2009) Proteomic analysis of increased Parkin expression and its interactants provides evidence for a role in modulation mitochondrial function. Proteomics 9:4284–4297

Dubinsky JM (2009) Heterogeneity of nervous system mitochondria: location, location, location! Exp Neurol 218:293–307

Duvaux-Miret O et al (1992) Immunosuppression in the definitive and intermediate hosts of the human parasite Schistosoma mansoni by release of immunoreactive neuropeptides. Proc Natl Acad Sci U S A 89:778–781
Eckman EA, Reed DK, Eckman CB (2001) Degradation of the Alzheimer's amyloid β peptide by endothelin-converting enzyme. J Biol Chem 276:24540–24548
Ecroyd H, Carver JA (2008) Unraveling the mysteries of protein folding and misfolding. IUBMB Life 60:769–774
Ehehalt R et al (2003) Amyloidogenic processing of the Alzheimer β-amyloid precursor protein depends on lipid rafts. J Cell Biol 160:113–123
El-Agnaf OM et al (2000) Oligomerization and toxicity of β-amyloid-42 implicated in Alzheimer's disease. Biochem Biophys Res Commun 273:1003–1007
Fountain JH, Kaur J, Lappin SL (2023) Physiology, renin angiotensin system. In: StatPearls. StatPearls Publishing, Treasure Island. https://www.ncbi.nlm.nih.gov/books/NBK470410/
Gan L, Cookson MR, Petrucelli L, La Spada AR (2018) Converging pathways in neurodegeneration, from genetics to mechanisms. Nat Neurosci 21:1300–1309
Garrido-Gil P, Joglar B, Rodriguez-Perez AI, Guerra MJ, Labandeira-Garcia JL (2012) Involvement of PPAR-γ in the neuroprotective and anti-inflammatory effects of angiotensin type 1 receptor inhibition: effects of the receptor antagonist telmisartan and receptor deletion in a mouse MPTP model of Parkinson's disease. J Neuroinflammation 9:38. https://doi.org/10.1186/1742-2094-9-38
Grimm MO, Mett J, Stahlmann CP, Haupenthal VJ, Zimmer VC, Hartmann T (2013) Neprilysin and Aβ clearance: impact of the APP intracellular domain in NEP regulation and implications in Alzheimer's disease. Front Aging Neurosci 5:98. Published 2013 Dec 23. https://doi.org/10.3389/fnagi.2013.00098
Guerreiro S, Privat AL, Bressac L, Toulorge D (2020) CD38 in neurodegeneration and neuroinflammation. Cells 9(2):471. Published 2020 Feb 18. https://doi.org/10.3390/cells9020471
Halliwell B (2006) Oxidative stress and neurodegeneration: where are we now? J Neurochem 97:1634–1658
Hauptmann S, Scherping I, Drose S et al (2009) Mitochondrial dysfunction: an early event in Alzheimer pathology accumulates with age in AD transgenic mice. Neurobiol Aging 30:1574–1586
Herring BP, El-Mounayri O, Gallagher PJ, Yin F, Zhou J (2006) Regulation of myosin light chain kinase and telokin expression in smooth muscle tissues. Am J Physiol Cell Physiol 291(5):C817–C827. https://doi.org/10.1152/ajpcell.00198.2006
Higashida H, Salmina AB, Olovyannikova RY, Hashii M, Yokoyama S, Koizumi K, Jin D, Liu H-X, Lopatina O, Amina S et al (2007) Cyclic ADP-ribose as a universal calcium signal molecule in the nervous system. Neurochem Int 51:192–199
Ho L et al (2004) Diet-induced insulin resistance promotes amyloidosis in a transgenic mouse model of Alzheimer's disease. FASEB J 18:902–904
Hutt DM, Powers ET, Balch WE (2009) The proteostasis boundary in misfolding diseases of membrane traffic. FEBS Lett 583:2639–2646
Iwata N et al (2000) Identification of the major Aβ1-42-degrading catabolic pathway in brain parenchyma: suppression leads to biochemical and pathological deposition. Nat Med 6:143–150
Iwata N et al (2004) Presynaptic localization of neprilysin contributes to efficient clearance of amyloid-β peptide in mouse brain. J Neurosci 24:991–998
Jellinger KA (2009) Recent advances in our understanding of neurodegeneration. J Neural Transm 116:1111–1162
Jellinger KA (2010) Basic mechanisms of neurodegeneration: a critical update. J Cell Mol Med 14(3):457–487. https://doi.org/10.1111/j.1582-4934.2010.01010.x
Kou W, Banerjee S, Eudy J, Smith LM, Persidsky R, Borgmann K, Wu L, Sakhuja N, Deshpande MS, Walseth TF et al (2009) CD38 regulation in activated astrocytes: implications for neuroinflammation and HIV-1 brain infection. J Neurosci Res 87:2326–2339
Koudinov AR, Berezov TT (2004) Alzheimer's amyloid-β (Aβ) is an essential synaptic protein, not neurotoxic junk. Acta Neurobiol Exp (Wars) 64:71–79

Kubo T, Yanagisawa H-a, Yagi T, Hirono M, Kamiya R (2010) Tubulin polyglutamylation regulates axonemal motility by modulating activities of inner-arm dyneins. Curr Biol 20:441–445
Lacroix B, van Dijk J, Gold ND, Guizetti J, Aldrian-Herrada G, Rogowski K, Gerlich DW, Janke C (2010) Tubulin polyglutamylation stimulates spastin-mediated microtubule severing. J Cell Biol 189:945–954
Lautrup S, Sinclair DA, Mattson MP, Fang EF (2019) NAD+ in brain aging and neurodegenerative disorders. Cell Metab 30:630–655
Ledesma MD, Da Silva JS, Crassaerts K, Delacourte A, De Strooper B, Dotti CG (2000) Brain plasmin enhances APP alpha-cleavage and Abeta degradation and is reduced in Alzheimer's disease brains. EMBO Rep 1(6):530–535. https://doi.org/10.1093/embo-reports/kvd107
Ledesma MD et al (2003) Proteomic characterisation of neuronal sphingolipid-cholesterol microdomains: role in plasminogen activation. Brain Res 987:107–116
Lehman NL (2009) The ubiquitin proteasome system in neuropathology. Acta Neuropathol 118:329–347
Leissring MA (2021) Insulin-degrading enzyme: paradoxes and possibilities. Cells 10(9):2445. https://doi.org/10.3390/cells10092445
Leissring MA et al (2003) Enhanced proteolysis of β-amyloid in APP transgenic mice prevents plaque formation, secondary pathology, and premature death. Neuron 40:1087–1093
Lichtenthaler SF, Haass C (2004) Amyloid at the cutting edge: activation of α-secretase prevents amyloidogenesis in an Alzheimer disease mouse model. J Clin Invest 113:1384–1387
Limoges A, Yarur HE, Tejeda HA (2022) Dynorphin/kappa opioid receptor system regulation on amygdaloid circuitry: implications for neuropsychiatric disorders. Front Syst Neurosci 16:963691. https://doi.org/10.3389/fnsys.2022.963691
Ling X et al (2002) Amyloid β antagonizes insulin promoted secretion of the amyloid β protein precursor. J Alzheimers Dis 4:369–374
Liu R et al (2004) Proteolytic antibody light chains alter β-amyloid aggregation and prevent cytotoxicity. Biochemistry 43:9999–10007
Mancuso M, Calsolaro V, Orsucci D et al (2009) Is there a primary role of the mitochondrial genome in Alzheimer's disease? J Bioenerg Biomembr 41:411–416
Marr RA et al (2003) Neprilysin gene transfer reduces human amyloid pathology in transgenic mice. J Neurosci 23:1992–1996
Martinet W, Agostinis P, Vanhoecke B et al (2009) Autophagy in disease: a double-edged sword with therapeutic potential. Clin Sci 116:697–712
Martucci LL, Amar M, Chaussenot R, Benet G, Bauer O, de Zélicourt A, Nosjean A, Launay J-M, Callebert J, Sebrié C et al (2019) A multiscale analysis in CD38 −/− mice unveils major prefrontal cortexdysfunctions. FASEB J 33:5823–5835
Migliore L, Coppede F (2009) Environmental-induced oxidative stress in neurodegenerative disorders and aging. Mutat Res 674:73–84
Miners S, Kehoe PG, Love S (2020) Cognitive impact of COVID-19: looking beyond the short term. Alzheimers Res Ther 1:170. https://doi.org/10.1186/s13195-020-00744-
Mirzaei H, Regnier F (2008) Protein:protein aggregation induced by protein oxidation. J Chromatogr B Analyt Technol Biomed Life Sci 873:8–14
Molnar Z et al (2004) Enhanced G-protein activation by a mixture of Aβ(25-35), Aβ(1-40/42) and zinc. J Neurochem 89:1215–1223
Morris AM, Watzky MA, Finke RG (2009) Protein aggregation kinetics, mechanism, and curve-fitting: a review of the literature. Biochim Biophys Acta 1794:375–397
Mullen RJ, Eicher EM, Sidman RL (1976) Purkinje cell degeneration, a new neurological mutation in the mouse. Proc Natl Acad Sci U S A 73:208–212
Nawrot B (2004) Targeting BACE with small inhibitory nucleic acids—a future for Alzheimer's disease therapy? Acta Biochim Pol 51:431–444
Nitsche A, Rose D, Fasold M, Reiche K, Stadler PF (2015) Comparison of splice sites reveals that long noncoding RNAs are evolutionarily well conserved. RNA 21:801
Nunomura A, Hofer T, Moreira PI et al (2009) RNA oxidation in Alzheimer disease and related neurodegenerative disorders. Acta Neuropathol 118:151–166

Palop JJ, Chin J, Mucke L (2006) A network dysfunction perspective on neurodegenerative diseases. Nature 443:768–773
Pang B, Araki A, Zhou L, Takebayashi H, Harada T, Kadota K (2023) CCP1, a regulator of tubulin post-translational modifications, potentially plays an essential role in cerebellar development. Int J Mol Sci 24(6):5335. Published 2023 Mar 10. https://doi.org/10.3390/ijms24065335
Perry TA, Greig NH (2004) A new Alzheimer's disease interventive strategy: GLP-1. Curr Drug Targets 5:565–571
Prince JA et al (2003) Genetic variation in a haplotype block spanning IDE influences Alzheimer disease. Hum Mutat 22:363–371
Prista von Bonhorst F, Gall D, Dupont G (2022) Impact of β-amyloids induced disruption of Ca2+ homeostasis in a simple model of neuronal activity. Cells 11(4):615. https://doi.org/10.3390/cells11040615
Quintana DS, Rokicki J, van der Meer D, Alnæs D, Kaufmann T, Córdova-Palomera A, Dieset I, Andreassen OA, Westlye LT (2019) Oxytocin pathway gene networks in the human brain. Nat Commun 10:668
Rangan SK et al (2003) Degradation of β-amyloid by proteolytic antibody light chains. Biochemistry 42:14328–14334
Regnard C, Desbruyeres E, Denoulet P, Eddé B (1999) Tubulin polyglutamylase: isozymic variants and regulation during the cell cycle in HeLa cells. J Cell Sci 112:4281–4289
Rogowski K, van Dijk J, Magiera MM et al (2010) A family of protein-deglutamylating enzymes associated with neurodegeneration. Cell 143(4):564–578. https://doi.org/10.1016/j.cell.2010.10.014
Ruse CI, Chin HG, Pradhan S (2022) Polyglutamylation: biology and analysis. Amino Acids 54(4):529–542. https://doi.org/10.1007/s00726-022-03146-4
Salman MM, Al-Obaidi Z, Kitchen P, Loreto A, Bill RM, Wade-Martins R (2021) Advances in applying computer-aided drug design for neurodegenerative diseases. Int J Mol Sci 22(9):4688. Published 2021 Apr 28. https://doi.org/10.3390/ijms22094688
Sambamurti K (2007) Secretases. In: Enna SJ, Bylund DB (eds) xPharm: the comprehensive pharmacology reference. Elsevier, Amsterdam, pp 1–5. https://doi.org/10.1016/B978-008055232-3.60569-0
Santiago TC, Parra L, Nani JV, Fidalgo TM, Bradshaw NJ, Hayashi MAF (2023) Angiotensin-converting enzymes as druggable features of psychiatric and neurodegenerative disorders. J Neurochem 166(2):138–155. https://doi.org/10.1111/jnc.15806
Savaskan E, Hock C, Olivieri G, Bruttel S, Rosenberg C, Hulette C, Müller-Spahn F (2001) Cortical alterations of angiotensin converting enzyme, angiotensin II and AT1 receptor in Alzheimer's dementia. Neurobiol Aging 4:541–546. https://doi.org/10.1016/s0197-4580(00)00259-1
Schmitz A et al (2004) Endoplasmic reticulum-localized amyloid β-peptide is degraded in the cytosol by two distinct degradation pathways. Traffic 5:89–101
Schroder M (2008) Endoplasmic reticulum stress responses. FASEB J 65:862–894
Schulman BA, Harper JW (2009) Ubiquitin-like protein activation by E1 enzymes: the apex for downstream signalling pathways. Nat Rev Mol Cell Biol 10:319–331
Schwartz AL, Ciechanover A (2009) Targeting proteins for destruction by the ubiquitin system: implications for human pathobiology. Annu Rev Pharmacol Toxicol 49:73–96
Seeley WW, Crawford RK, Zhou J et al (2009) Neurodegenerative diseases target largescale human brain networks. Neuron 62:42–52
Shirotani K et al (2001) Neprilysin degrades both amyloid β peptides 1-40 and 1-42 most rapidly and efficiently among thiorphan- and phosphoramidon-sensitive endopeptidases. J Biol Chem 276:21895–21901
Smith MA, Zhu X, Tabaton M et al (2010) Increased iron and free radical generation in preclinical Alzheimer disease and mild cognitive impairment. J Alzheimers Dis 19:363–372
Sonntag K-C, Ryu W-I, Amirault KM, Healy RA, Siegel AJ, McPhie DL, Forester B, Cohen BM (2017) Late-onset Alzheimer's disease is associated with inherent changes in bioenergetics profiles. Sci Rep 7:14038

Sultana R, Butterfield DA (2009) Oxidatively modified, mitochondria-relevant brain proteins in subjects with Alzheimer disease and mild cognitive impairment. J Bioenerg Biomembr 41:441–446

Suryavanshi S, Edde B, Fox LA, Guerrero S, Hard R, Hennessey T, Kabi A, Malison D, Pennock D, Sale WS et al (2010) Tubulin glutamylation regulates ciliary motility by altering inner dynein arm activity. Curr Biol 20:435–440

Swerdlow RH, Khan SM (2009) The Alzheimer's disease mitochondrial cascade hypothesis: an update. Exp Neurol 218:308–315

Swerdlow RH, Burns JM, Khan SM (2014) The Alzheimer's disease mitochondrial cascade hypothesis: progress and perspectives. Biochim Biophys Acta 1842(8):1219–1231. ISSN 09254439. https://doi.org/10.1016/j.bbadis.2013.09.010

Tai HC, Schuman EM (2008) Ubiquitin, the proteasome and protein degradation in neuronal function and dysfunction. Nat Rev Neurosci 9:826–838

Tanco S, Tort O, Demol H et al (2015) C-terminomics screen for natural substrates of cytosolic carboxypeptidase 1 reveals processing of acidic protein C termini. Mol Cell Proteomics 14(1):177–190. https://doi.org/10.1074/mcp.M114.040360

Thomas J, Smith H, Smith CA, Coward L, Gorman G, De Luca M, Jumbo-Lucioni P (2021) The angiotensin-converting enzyme inhibitor lisinopril mitigates memory and motor deficits in a drosophila model of Alzheimer's disease. Pathophysiology 2:307–319. https://doi.org/10.3390/pathophysiology28020020

Torika N, Asraf K, Roasso E, Danon A, Fleisher-Berkovich S (2016) Angiotensin converting enzyme inhibitors ameliorate brain inflammation associated with microglial activation: possible implications for Alzheimer's disease. J Neuroimmune Pharmacol 4:774–785. https://doi.org/10.1007/s11481-016-9703-8

Torruella-Suarez ML, McElligott ZA (2020) Neurotensin in reward processes. Neuropharmacology 167:108005

Tschumi CW, Blankenship HE, Sharma R, Lynch WB, Beckstead MJ (2022) Neurotensin release from dopamine neurons drives long-term depression of substantia nigra dopamine signaling. J Neurosci 42(32):6186–6194. https://doi.org/10.1523/JNEUROSCI.1395-20.2022

Tucker HM et al (2000) Tissue plasminogen activator requires plasminogen to modulate amyloid-β neurotoxicity and deposition. J Neurochem 75:2172–2177

Turner AJ, Tanzawa K (1997) Mammalian membrane metallopeptidases: NEP, ECE, KELL, and PEX. FASEB J 11:355–364

Wakade C, Chong R, Bradley E, Thomas B, Morgan J (2014) Upregulation of GPR109A in Parkinson's disease. PLoS One 9:e109818

Wang T, Morgan JI (2007) The Purkinje cell degeneration (pcd) mouse: an unexpected molecular link between neuronal degeneration and regeneration. Brain Res 1140:26–40

Wang S, Wang R, Chen L, Bennett DA, Dickson DW, Wang DS (2010) Expression and functional profiling of neprilysin, insulin-degrading enzyme, and endothelin-converting enzyme in prospectively studied elderly and Alzheimer's brain. J Neurochem 115(1):47–57. https://doi.org/10.1111/j.1471-4159.2010.06899.x

Wang X, Zhang Q, Bao R, Zhang N, Wang Y, Polo-Parada L, Tarim A, Alemifar A, Han X, Wilkins HM et al (2017) Deletion of Nampt in projection neurons of adult mice leads to motor dysfunction, neurodegeneration, and death. Cell Rep 20:2184–2200

Werb Z (1997) ECM and cell surface proteolysis: regulating cellular ecology. Cell 91:439–442

Winklhofer KF, Tatzelt J, Haass C (2008) The two faces of protein misfolding: gain- and loss-of-function in neurodegenerative diseases. EMBO J 27:336–349

Xie L et al (2002) Alzheimer's β-amyloid peptides compete for insulin binding to the insulin receptor. J Neurosci 22:RC221

Yang Y, Lu B (2009) Mitochondrial morphogenesis, distribution, and Parkinson disease: insights from PINK1. J Neuropathol Exp Neurol 68:953–963

Yang JL, Weissman L, Bohr VA et al (2008) Mitochondrial DNA damage and repair in neurodegenerative disorders. DNA Repair 7:1110–1120

Yasojima K et al (2001) Reduced neprilysin in high plaque areas of Alzheimer brain: a possible relationship to deficient degradation of β-amyloid peptide. Neurosci Lett 297:97–100

Ye Y, Rape M (2009) Building ubiquitin chains: E2 enzymes at work. Nat Rev Mol Cell Biol 10:755–764

Role of Enzymes in Alzheimer's Disease

2

Sumitra Patel, Shweta Shrivastava, Parameswary V. Marar, Rajan Swami, and Manish Kumar Jeengar

Abstract

This chapter aims to explain the important roles that specific enzymes play in the etiology of Alzheimer's Disease (AD). Emphasis is placed on enzymes such as beta-secretase 1, gamma-secretase, and insulin-degrading enzyme (IDE), which significantly contribute to the formation of hallmark pathological features of AD—the amyloid beta plaques and Tau protein aggregates. Furthermore, this chapter critically evaluates the latest advancements in the study of enzyme inhibitors as potential therapeutic agents. It underscores the promise they hold in decelerating or even halting the progression of AD while simultaneously addressing the challenges encountered in the design and development of such inhibitors.

By delving into the complexity of enzyme function and their manipulation, this chapter enriches our ongoing discourse on the exploration of promising treatments for neurodegenerative diseases. The aim is not only to deepen our understanding of the biochemical underpinnings of AD but also to highlight the potential of targeted enzyme intervention as a novel approach in the fight against this debilitating disease.

S. Patel · P. V. Marar · M. K. Jeengar (✉)
Department of Pharmacology, Amrita School of Pharmacy, Amrita Vishwa Vidyapeetham, AIMS Health Sciences Campus, Kochi, Kerala, India
e-mail: manishkumarj@pharmacy.aims.amrita.edu

S. Shrivastava
School of Pharmacy, School of Health and Allied Sciences, ARKA JAIN University, Jamshedpur, Jharkhand, India

R. Swami
Chitkara College of Pharmacy, Chitkara University, Rajpura, Punjab, India

B. Mathew, D. G. T. Parambi (eds.), *Enzymes in Neurodegenerative Disorders*,
https://doi.org/10.1007/978-981-97-6822-6_2

Keywords

Alzheimer's disease · β-Secretase (BACE1) · Acetylcholinesterase · γ-Secretase substrates · Neprilysin · Insulin-degrading enzyme (IDE)

2.1 Introduction

As the most prevalent form of dementia, Alzheimer's disease (AD) presents a critical global health issue, and it is anticipated that the number of individuals affected by this disease will increase to 152 million by 2050 (Soria Lopez et al. 2019). Alarmingly, this number is projected to triple by the middle of the century due to the global rise in life expectancy. AD is a neurodegenerative condition marked by a gradual decline in cognitive abilities and memory retention (Fig. 2.1). Over time, these symptoms progress to a severe impairment of daily functions. The disease derives its name from Dr. Alois Alzheimer, who, in 1906, first identified "senile plaques" and "neurofibrillary tangles (NFTs)" in the brain of a patient posthumously, who had demonstrated signs of dementia.

The distinguishing features of AD—extracellular amyloid-beta (Aβ) plaques and intracellular NFTs of hyperphosphorylated tau protein—emerge from the complex interaction of genetic, environmental, and lifestyle factors (Breijyeh and Karaman 2020; Oddo et al. 2003). Although the exact trigger for AD remains elusive, the prevalent amyloid cascade hypothesis proposes that the accumulation of Aβ peptides

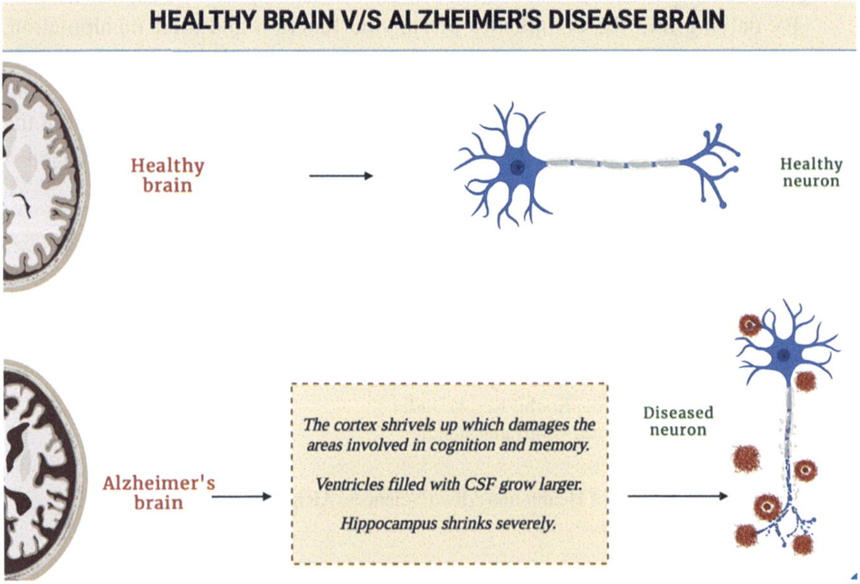

Fig. 2.1 The physiological structure of the brain and neurons in a healthy brain and Alzheimer's brain

initiates the AD pathogenesis, catalyzing subsequent events, such as tau hyperphosphorylation, inflammation, oxidative stress, and culminating in neuronal death.

A host of key enzymes, including beta-secretase 1 (BACE1) and gamma-secretase, are implicated in AD's pathogenesis as shown in Fig. 2.2. These enzymes participate in the cleavage of the amyloid precursor protein (APP), leading to the production of Aβ peptides. On the other side, the insulin-degrading enzyme (IDE) primarily contributes to the degradation of Aβ peptides. Consequently, dysfunctions or modifications in these and other pertinent enzymes can tip the balance between Aβ production and clearance, fueling Aβ accumulation and plaque formation.

The risk of developing AD is significantly steered by genetic factors. Familial early-onset AD, which accounts for a minor fraction of all AD cases, is commonly linked to mutations in the APP, PSEN1, or PSEN2 genes. In contrast, late-onset AD, the more prevalent form, is influenced by a broader genetic landscape, with numerous genes contributing to the disease risk. The ε4 variant of the APOE gene stands out as the most significant genetic risk factor for late-onset AD (Silva et al. 2014). Additional genetic variants, particularly those in genes encoding for enzymes involved in Aβ metabolism, can also shape the AD risk and progression.

Despite the relentless pursuit of a cure, AD remains an incurable disease. Present treatments primarily focus on symptom management and decelerating disease progression. In recent years, attention has been focused on the development of several techniques that target the AChE enzyme as well as other targets such butyl cholinesterase (BChE), amyloid-β(Aβ), β-secretase-1 (BACE), metals antioxidant, and

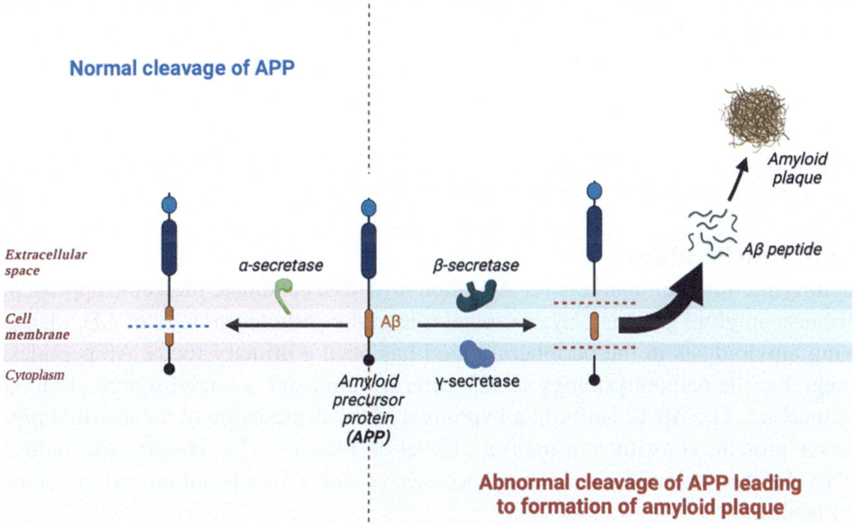

Fig. 2.2 Role of secretase enzymes in amyloid precursor protein (APP) processing. The cleavage process APP can occur in normal and pathological states. When in normal conditions, APP is first cleaved by α-secretase, which generates a sAPP product. However, in pathological conditions, APP is cleaved sequentially by β-secretase and γ-secretase, which produces an Aβ42 fragment that tends to aggregate and form plaques. Ultimately, this leads to cell death

free radical scavenging ability. However, the in-depth understanding of enzymes' role in AD pathogenesis provides promising leads for innovative therapeutic strategies. By manipulating these enzymes, there exists a possibility to modulate Aβ metabolism and other vital pathological processes, potentially transforming the trajectory of this devastating disease.

This chapter aims to offer a thorough exploration of enzymes' role in AD, with a particular emphasis on BACE1, gamma-secretase, and IDE. We will delve into their biological functions, structures, the repercussions of their dysfunction on AD, potential genetic modifiers, and how these enzymes can be leveraged for treatment. We aspire that this discourse will broaden our comprehension of AD and add momentum to the quest for efficacious therapies.

2.1.1 Biochemistry and Molecular Biology of AD

Early diagnosis of AD through biomarkers could significantly alter the disease's course by enabling early therapeutic measures and monitoring. Traditional AD biomarkers, including core cerebrospinal fluid (CSF) biomarkers like amyloid beta (Aβ), total tau, phosphorylated tau protein, and positron emission tomography (PET) imaging of glucose metabolism and amyloid deposition, have limitations due to their invasiveness and high cost.

As an alternative, blood biomarkers emerge as a more time-efficient and cost-effective solution. However, given the progressive nature of AD and uncertainties surrounding the degree of blood-brain barrier (BBB) integrity loss, the identification of reliable blood biomarkers for AD remains challenging. Research suggests that BBB disruption in AD patients correlates with the exchange of proteins and other molecules between the CSF and blood. Additionally, the complexity of blood as a biological fluid and its many confounding factors necessitate strict adherence to standardized protocols in the collection and analysis of blood samples (Altuna-Azkargorta and Mendioroz-Iriarte 2018).

2.1.1.1 Aβ Peptides

Identifying blood biomarkers for AD often utilizes a candidate molecule approach. Because amyloid genesis plays a crucial role in the pathophysiology of AD, identifying amyloidosis in the peripheral blood has been a primary focus. Aβ peptides, integral to the pathophysiology of AD, offer a significant potential source of blood biomarkers. The Aβ 42 isoform, a byproduct of the degradation of the amyloid precursor protein, constitutes a majority of senile plaques. This isoform can induce DNA damage via oxidative stress processes, posing a significant hazard to neuronal health.

Circulating Aβ originates from both peripheral and central sources, and the relationship between Aβ in the blood and CSF is quite nuanced. An intriguing observation is the reportedly higher serum Aβ1-42 levels in AD patients. In older populations, elevated plasma Aβ 42 levels, low Aβ 40 levels, and reduced Aβ 42/Aβ 40 ratios may indicate a cognitive shift toward mild cognitive impairment (MCI) or AD (Altuna-Azkargorta and Mendioroz-Iriarte 2018).

2.1.1.2 Enzymes Involved in Tau Protein Phosphorylation

Investigations have also been directed toward enzymes involved in tau protein phosphorylation. The phosphorylation of tau protein plays a critical role in the formation of NFTs, a definitive sign of AD. This phosphorylation regulates the tau protein's ability to form oligomers and aggregates. Ultra-sensitive immunoanalytical techniques have revealed an increase in tau protein levels in AD patients. Certain protein kinases such as glycogen synthase kinase 3 (GSK-3) and dual specificity tyrosine-phosphorylation regulated kinase A (DYRK1A) have been implicated in tau protein hyperphosphorylation, suggesting their potential as biomarkers (Altuna-Azkargorta and Mendioroz-Iriarte 2018).

2.1.1.3 Neurofilament Light Polypeptide

Neurofilament light polypeptide (NF-L), a marker of neuronal damage, is another significant player in the search for blood biomarkers. NF-L concentrations are found to be high in the CSF of AD patients, and increased plasma NF-L levels are reported in Aβ-positive patients with AD and MCI (Altuna-Azkargorta and Mendioroz-Iriarte 2018).

2.1.1.4 Inflammatory Biomarkers

In line with the inflammatory hypothesis of AD, which postulates that neuroinflammation contributes to the disease's early stages and progression, inflammatory mediators like cytokines and chemokines present in the CSF and blood may also serve as early diagnostic biomarkers for AD. Elevated levels of certain cytokines, such as interleukins (IL-1, IL-4, IL-6, IL-10), interferon, and tumor necrosis factor α (TNF-α), are found in the brains of AD patients. The glycoprotein cytokine I-309, released by activated T cells, has been proposed as a potential marker of MCI's transition to AD (Altuna-Azkargorta and Mendioroz-Iriarte 2018).

2.1.1.5 Apoptosis Biomarkers and Autoantibodies

Apoptosis biomarkers like clusterin, involved in neurodegenerative processes, are found in higher concentrations in the blood of AD patients. Also, a decrease in levels of certain redox-reactive antiphospholipid autoantibodies in the CSF of AD patients has been reported, providing potential diagnostic markers for early disease stages.

2.1.1.6 Platelet Biomarkers and Microvascular Damage Biomarkers

Platelets and brains of AD patients demonstrate increased B-secretase activity, monoamine oxidase b expression, and phospholipase A2 activity, presenting another potential biomarker route. Meanwhile, markers indicating microvascular damage, such as elevated levels of atrial natriuretic peptide and adrenomedullin, have been detected in the prodromal stages of AD.

2.1.1.7 Homocysteine and Cystatin C

The role of homocysteine, an amino acid linked to both vascular dementia and AD, has come into focus as a potential biomarker. It has been observed that plasma homocysteine levels appear to have a close relationship with Aβ42 levels. Furthermore, AD patients exhibit significantly increased plasma homocysteine levels, making it a potential biomarker of interest. The metabolism of homocysteine is reliant on folic acid, which regulates DNA methyltransferase, reducing Aβ synthesis.

Finally, cystatin C, a cysteine protease inhibitor that suppresses Aβ aggregation and deposition by binding to the amyloid precursor protein and Aβ40 and Aβ42 peptides, has been found in reduced levels in the serum and CSF of AD patients (Altuna-Azkargorta and Mendioroz-Iriarte 2018). Table 2.1 provides a summary of potential biomarkers for AD diagnosis.

2.2 Enzymes Implicated in AD

Enzymes serve as significant molecular machines in the progression of AD, instigating complex biological processes that lead to the characteristic symptoms of this devastating condition. Several key enzymes contribute to the formation of pathological structures seen in AD, such as amyloid-beta plaques and NFTs. In this section, we delve into the roles and implications of some central enzymes—BACE1, gamma-secretase, IDE, and tau kinases, each providing a unique perspective on AD pathogenesis. Table 2.2 summarizes the information about the main enzymes implicated in AD.

2.2.1 BACE1

Among the defining pathological features of AD is the accumulation and aggregation of Aβ peptides into senile plaques, predominantly facilitated by the enzymatic

Table 2.1 A list of biomarkers with potential for use in AD diagnosis

Molecule	Level	Pathogenic pathway
Apolipoprotein A1	Decreased in AD and MCI	Amyloidogenic
Apolipoprotein A4	Increased in AD and MCI	Vascular damage
Alpha-2-HS-glycoprotein	Decreased in AD and MCI	Inflammatory
Afamin	Decreased in AD	Amyloidogenic
α-1-antitrypsin	Increased in AD	Amyloidogenic
α-2-macroglobulin	Increased in AD	Amyloidogenic
Complement C3	Increased in AD	Inflammatory
Fibrinogen gamma chain	Increased in AD and MCI	Vascular damage
Plasminogen	Decreased in AD	Vascular damage

AD Alzheimer's disease, *MCI* mild cognitive impairment

Table 2.2 Key enzymes implicated in Alzheimer's disease

Enzyme	Biological role in AD pathogenesis	Protein structure and interacting domains	Dysfunction impact on AD	Potential genetic modifiers	References
Beta-secretase 1 (BACE1)	Cleaves amyloid precursor protein (APP) to form amyloid beta (Aβ) peptides	Transmembrane aspartyl protease with catalytic domain influencing the protease activity	Overexpression or increased activity leads to elevated levels of Aβ peptides and potential plaque formation	Genetic variants such as rs638405 could influence BACE1 function and Aβ production	Bazzari and Bazzari (2022)
Gamma-secretase	Cleaves the transmembrane region of APP (C-terminal transmembrane fragment) to produce Aβ	Intramembrane aspartyl protease that cleaves its substrates within their transmembrane domains	Increased activity leads to elevated levels of Aβ peptides and the formation of plaque	Genetic variants of *TREM2* (ex. R47H) leads to accumulation of substrate-C-terminal fragments	Sathya et al. (2012)
Alpha secretase	Cleaves Aβ precursor protein (AβPP) preventing the production and aggregation of Aβ	Proteolytic enzymes that cleave APP in its transmembrane region	Overexpression increases the levels of Aβ	Modulating catalytic activity of ADAM 10	Ahmad et al. (2019)
Insulin degrading enzyme	Degradation and clearance of Aβ	Zinc metalloprotease that degrades soluble, monomeric Aβ and is localized in the cytoplasm as well as on the cell surface and within the mitochondria	Decreased levels lead to the accumulation of Aβ	Single nucleotide polymorphism rs2421943 increases the risk of AD	Taylor et al. (2022)
Neprilysin	β-Amyloid (Aβ)-degrading enzyme	Zinc-dependent membrane metalloprotease	Downregulation leads to the accumulation of β-amyloid	rs9827586 was significantly associated with NEP protein level	Alam and Sharma (2018)

activities of β-secretase and γ-secretase (Mancini et al. 2011). Of particular importance in this process is the enzyme BACE1, also referred to as β-site APP cleaving enzyme 1. As a membrane-bound aspartyl protease, BACE1 bears similarities to the pepsin and retroviral aspartic protease families (Yu et al. 2016). Additionally, it is known by the designations of asp2 or memapsin2 (Mancini et al. 2011; Alam and Sharma 2019). BACE1 plays a pivotal role in the genesis of the amyloidogenic peptide Aβ42 (42-residue Aβ peptide) (Altuna-Azkargorta and Mendioroz-Iriarte 2018), the key component of Aβ plaques. Notably, in patients with late-onset AD (LOAD), there is a reported increase in BACE1 concentrations in the brain and CSF (Cervellati et al. 2020).

The proteolytic process of APP, a large, ubiquitously expressed, type 1 integral protein, results in the production of Aβ peptide. This process involves the pro-amyloidogenic activities of β- and γ-secretase, and the anti-amyloidogenic actions of α-secretase (Cole and Vassar 2008). As the initial enzyme in the amyloidogenic process, BACE1 cleaves APP to yield the N-terminus, with γ-secretase subsequently cleaving to form the C-terminus. This sequence of actions culminates in the formation of the Aβ peptide (Mancini et al. 2011). The brain tissue of sporadic AD patients has shown a correlation between heightened BACE1 activity and Aβ peptide burden. Intriguingly, BACE1 exhibits a preference for a low pH environment and is primarily localized in acidic intracellular compartments such as endosomes and the trans-Golgi network (Cole and Vassar 2008).

APP undergoes processing via two primary pathways: the amyloidogenic and non-amyloidogenic pathways. In the non-amyloidogenic pathway, APP is initially cleaved by α-secretase at amino acid 687 (in the APP 770 isoform), releasing the soluble APP peptide (sAPPα) ectodomain into the extracellular space. Concurrently, a C-terminal fragment (CTF) that is 83 amino acids in length, C83, remains within the plasma membrane. Subsequent cleavage by γ-secretase results in a non-toxic or non-amyloidogenic 3 kDa fragment, p3, which is released into the extracellular space, and the APP intracellular domain (ICD) that is deposited into the cytoplasm (Fig. 2.3). The α-secretase activity is attributed to three proteases: TACE (TNF-α converting enzyme), ADAM-9 (a disintegrin and metalloprotease domain protein-9), and ADAM-10.

Turning our attention to the amyloidogenic pathway, APP is first cleaved by β-secretase at amino acid 671 (in the APP 770 isoform). This cleavage forms the N-terminus of Aβ at the Asp+1 residue in the Aβ sequence, thereby producing a soluble fragment known as sAPPβ that is released into the extracellular space. This cleavage results in two segments: a secreted APP ectodomain, APPsβ, and a membrane-bound CTF that is 99 amino acids in length, C99.

The γ-secretase complex, comprising PS1 or PS2, nicastrin, Aph1, and Pen2, processes the remaining 99 amino acid CTF of APP (C99) into Aβ40/42 within endocytic components. Most of the cleavage concludes at amino acid 40 (Aβ40), and a smaller portion terminates at amino acid 42 (Aβ42). These fragments are then released into the extracellular space and the cytoplasm as Aβ and AICD, respectively. It is noteworthy that Aβ42, having a higher tendency to aggregate, constitutes the majority of the amyloid plaques compared to Aβ40 (Cole and Vassar 2008; De

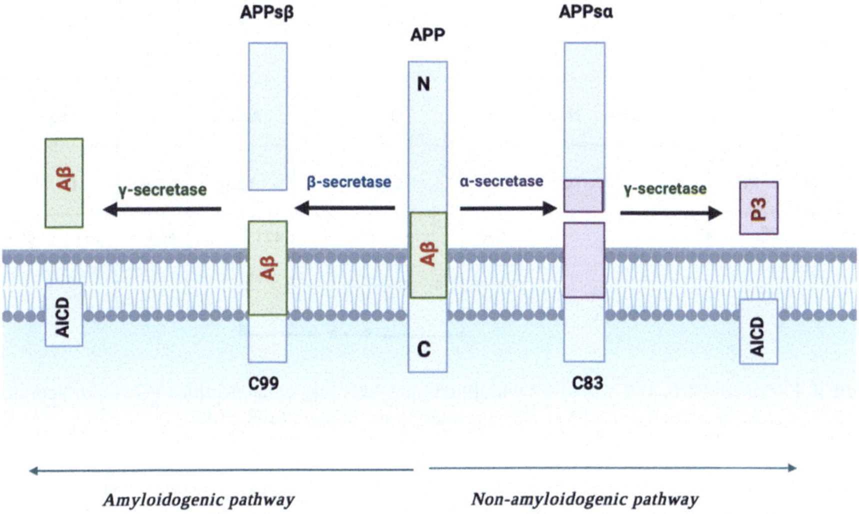

Fig. 2.3 The amyloid precursor protein (APP) goes through two different pathways for processing, the non-amyloidogenic and the amyloidogenic pathway. The APP can be cleaved by either α- or β-secretase. When cleaved by α-secretase (the non-amyloidogenic pathway), it generates two fragments, sAPP-α and C83. When cleaved by β-secretase (the amyloidogenic pathway), it generates two different fragments, sAPP-β and C99. C83 is then cleaved by γ-secretase, which generates AICD and p3. Similarly, C99 is also cleaved by γ-secretase, which generates AICD and Aβ

Strooper et al. 2010; Hur 2022; Tolia and De Strooper 2009). Familial Alzheimer's disease (FAD) mutations impact γ-secretase-dependent cleavage, leading to an overproduction of Aβ42. Often, an increase in non-amyloidogenic APP metabolism accompanies a reduction in the amyloidogenic pathway, and vice versa, as the enzymes, α- and β-secretase, compete for the APP substrate. Hence, β-secretase, being the rate-limiting step, is considered a primary drug target to reduce Aβ levels in AD (Cole and Vassar 2008).

2.2.1.1 Structure of BACE1

Beta-site amyloid precursor protein cleaving enzyme 1 (BACE1), a crucial enzyme in the pathogenesis of AD, exhibits a complex structural conformation that is vital for its function. The enzyme is a type I transmembrane aspartyl protease, closely related to pepsin, cathepsin, and renin, all belonging to the pepsin family A1 of aspartyl proteases.

BACE1 consists of a signal peptide, a prodomain, a catalytic domain, a transmembrane domain, and a C-terminal domain. It is synthesized as a proenzyme and undergoes a series of post-translational modifications, including glycosylation and proteolytic cleavage, to become mature and active. The catalytic domain of BACE1 consists of two aspartic acid residues (Asp93 and Asp289) situated within two lobes that form a cleft where substrates bind (Fig. 2.4). This cleft has been of particular interest because its blockage could potentially stop the enzyme from

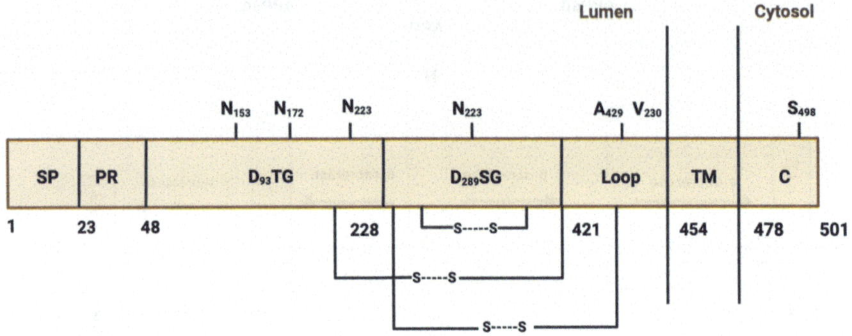

Fig. 2.4 Primary structure of BACE1 with functional domains: signal peptide (1–23), pro-peptide (23–27), catalytic domain (47–454), and transmembrane domain (478–501)

producing toxic beta-amyloid fragments. A series of loops, termed flap regions, control access to this active site. The transmembrane domain anchors BACE1 to cell membranes, predominantly in intracellular compartments like the Golgi apparatus and endosomes, where the pH is favorable for its activity. The C-terminal domain, on the other hand, is implicated in the sorting and trafficking of BACE1 within cells. The overall three-dimensional structure of BACE1 reveals a homodimeric arrangement, although it is still debated whether this dimerization is required for its function. Nonetheless, the architecture of BACE1 provides significant insights into the design of specific inhibitors aimed at its active site to prevent the formation of amyloid-beta peptides. Understanding the detailed structure of BACE1 aids in the design of therapeutic strategies for AD. The enzyme's topography presents multiple potential sites for intervention by small molecule inhibitors or biological agents, opening avenues for innovative treatment approaches against AD.

2.2.1.2 Biological Role

BACE1, recognized for its role in β-amyloid production—a characteristic trait of AD—is highly expressed in the brain. However, BACE1 mRNA and protein are also present in a myriad of other cell types and tissues, including adipocytes, hepatocytes, vascular cells, bone marrow, lymphoid and respiratory tissues, muscle tissue, and pancreatic β-cells. Elevated BACE1 levels are pathologically associated with the development of metabolic diseases such as type 2 diabetes, cardiovascular diseases, and obesity. A significant portion of research on BACE1 has been dedicated to understanding its role in the amyloidogenic pathway, specifically its contribution to the generation of Aβ40 and Aβ42 peptides via sequential cleavage of the APP protein. BACE2, despite sharing approximately 64% homology with BACE1, cleaves the APP protein at a different site and does not elicit the same β-secretase action. It is primarily found in peripheral tissues, while BACE1 is predominantly located in the brain. Interestingly, BACE1 also plays a role in regulating myelin repair, synaptic function, and memory. The accumulation of Aβ plaques can disrupt synaptic functions, leading to cognitive impairments, cerebrovascular complications, BBB impairments, vascular injuries, and capillary degradation. Moreover, Aβ has been discovered to regulate the transcription of the insulin-like growth factor receptor, further underscoring its complex role in biological systems (Lauer et al. 2020).

2.2.1.3 Dysfunction of BACE1 in AD

Beta-site APP cleaving enzyme 1 (BACE1) plays a pivotal role in AD pathology due to its essential involvement in the production of amyloid-beta (Aβ) peptides. A dysfunction in the normal activity of this enzyme can trigger a series of events leading to the characteristic features of AD. Overexpression of BACE1 is frequently observed in the brains of AD patients, and this increase in enzyme activity correlates with a higher load of Aβ peptides, contributing to plaque formation. Elevated BACE1 activity has been linked with both sporadic and familial AD, emphasizing its role in different forms of the disease. Moreover, BACE1 dysregulation not only influences the amyloidogenic pathway but also affects various physiological processes due to its involvement in the cleavage of other substrates. These include proteins involved in myelination, synaptic activity, and axonal growth, which are all critical for proper brain function. As such, increased BACE1 activity in AD could have far-reaching consequences beyond Aβ production. Several factors can contribute to BACE1 dysfunction in AD. Chronic stress, oxidative damage, and inflammation, common in AD, have all been shown to increase BACE1 expression and activity. Additionally, alterations in BACE1 trafficking and localization within the cell, which could be influenced by genetic factors or cellular stress, can affect its activity and result in enhanced production of Aβ peptides.

The multifaceted role of BACE1 in AD underlines the importance of understanding its normal function and how its dysregulation can contribute to disease pathogenesis. This knowledge is crucial for the development of therapeutic strategies aimed at modulating BACE1 activity or function.

2.2.2 Gamma-Secretase

AD manifests in two primary forms: sporadic and familial. The familial type is attributed to mutations in the presenilin (PS) genes, specifically presenilin 1 (PS1) and presenilin 2 (PS2). These mutations lead to the formation of Aβ peptides from the APP protein, accumulating in the form of senile plaques or amyloid, characteristic markers of the disease. γ-Secretase, a critical enzyme in this process, is a multi-subunit aspartyl protease responsible for the final cleavage of the APP, an integral step in the metabolic pathway of AD. This enzyme is classified under intramembrane-cleaving proteases (I-CliPs), a novel class of proteases (Miners et al. 2008; Sousa et al. 2021).

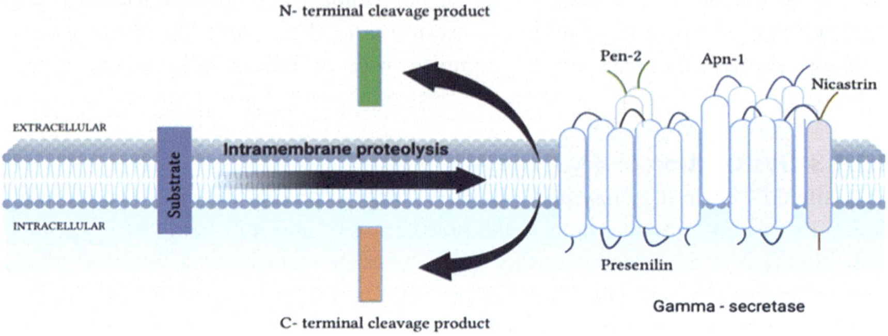

Fig. 2.5 γ-Secretase stands out as a multi-subunit enzyme. Presenilin (PS) serves as the primary contributor, endowing it with catalytic prowess. The complex also houses accessory integral membrane proteins, such as nicastrin (NCT), anterior-pharynx defective 1 (Aph-1), and PS-enhancer 2 (PEN-2), orchestrating the comprehensive functioning of the entire complex

2.2.2.1 Structure

The γ-secretase complex comprises multiple proteins, each playing a distinct role. The principal parts, which impart catalytic activity to the complex, are PS1 and PS2. Accessory proteins, like nicastrin, anterior pharynx-defective 1 (APH-1), and presenilin enhancer 2, contribute to the overall function and stability of the γ-secretase complex (Fig. 2.5). For instance, nicastrin serves as a substrate receptor (Miners et al. 2008; Sousa et al. 2021) PS1 and PS2, integral to the amyloidogenic processing of APP, possess the enzyme's active site situated at their aspartyl residues. Interestingly, mature PS undergoes cleavage between its sixth and seventh transmembrane domains, leading to the formation of NTF and CTF. The resultant heterodimer, consisting of PS-NTF and PS1-CTF, represents the catalytically active form of γ-secretase (Kurochkin et al. 2018).

2.2.2.2 Biological Role

γ-Secretase plays a pivotal role in the cleavage of APP. Depending on the cleavage pathway—amyloidogenic or non-amyloidogenic—the outcomes vary. In the non-amyloidogenic pathway, α-secretase first cleaves APP, leading to aAPPα and APP-CTF products. Subsequent cleavage by γ-secretase results in a non-pathologic, non-neurotoxic amyloid peptide termed P3. Contrarily, in the amyloidogenic route, β-secretase (BACE1) acts on APP, generating sAPPβ and the membrane-bound APP-CTF (C99). C99 then undergoes γ-secretase-mediated cleavage, culminating in the production of Aβ and the APP intracellular domain (AICD), which translocates to the nucleus. Distinguishing between the cleavage sites, γ-secretase has three sites termed γ-, ζ-, and ϵ-sites. Aβ42, originating from the γ-site, is particularly significant as it is more prone to aggregation and toxicity compared to Aβ40. Given its central role in the pathogenesis of AD, γ-Secretase has emerged as a vital target for therapeutic intervention. Yet, challenges persist, especially given the enzyme's subunit diversity, which necessitates selective subunit targeting over broad inhibition (Miners et al. 2008; Kurochkin et al. 2018).

2.2.2.3 Dysfunction

The ability of γ-secretase to mediate membrane-to-nucleus signaling through regulated intramembrane proteolysis is crucial for its function. Its unique ability allows it to cleave peptide bonds within the membrane's hydrophobic environment. However, dysfunction arises with certain mutations, especially in the PSEN1 and PSEN2 genes. Over 150 mutations in PSEN1 and about 10 in PSEN2 are associated with early-onset FAD. Many of these mutations lie within hydrophobic domains and lead to an altered ratio of amyloidogenic Aβ42 peptide relative to Aβ40. These mutations might not only cause a loss of function by altering substrate binding but may also impact the binding affinity for inhibitors and modulators (Qiu and Folstein 2006).

2.2.3 IDE

The equilibrium of Aβ levels in the body is maintained by a delicate balance determined by factors such as Aβ production, clearance, and degradation. Integral to this balance are proteases like neprilysin (NEP) and IDE (Chen et al. 2021). IDE, also denoted as insulysin, insulinase, or insulin protease, is a ubiquitous zinc metalloendopeptidase enzyme with its presence predominantly observed in tissues like the liver, testis, muscles, and brain. The enzyme, weighing around 110 kDa, is encoded by the IDE gene located on the chromosome band 10q23-q25 (Nalivaeva et al. 2020). Its multifaceted functions involve the degradation of a spectrum of substrates ranging from insulin, Aβ-peptides, islet amyloid polypeptide (IAPP or amylin), glucagon, to insulin-like growth factors I and II (Grimmer et al. 2019). In the context of AD, IDE's role in the degradation of Aβ-peptides is particularly critical.

Beyond its well-known proteolytic functions, IDE assumes diverse roles as a chaperone/dead-end chaperone, protease modulator, E1-ubiquitin activating enzyme, and a key player in the heat shock response mechanism (Chen et al. 2021; Sharma et al. 2021).

2.2.3.1 Structure

The human IDE structure is intricate, with its gene comprising 25 exons. Among its six transcripts, four contain the catalytic site sequence, and the remaining two encompass the sequence for the active site. Its architecture is a marvel of symmetry, possessing four homologous domains. The first two domains, denoted as domains 1 and 2, make up the N-terminal (IDE-N). In contrast, domains 3 and 4 constitute the C-terminal (IDE-C). These domains are linked by a 28-residue loop, forming a unique prism-shaped cavity. The enzyme, in its closed conformation, restricts substrate binding. However, specific conformational changes facilitate the substrate to bind and consequently initiate its action (Fig. 2.6) (Grimmer et al. 2019).

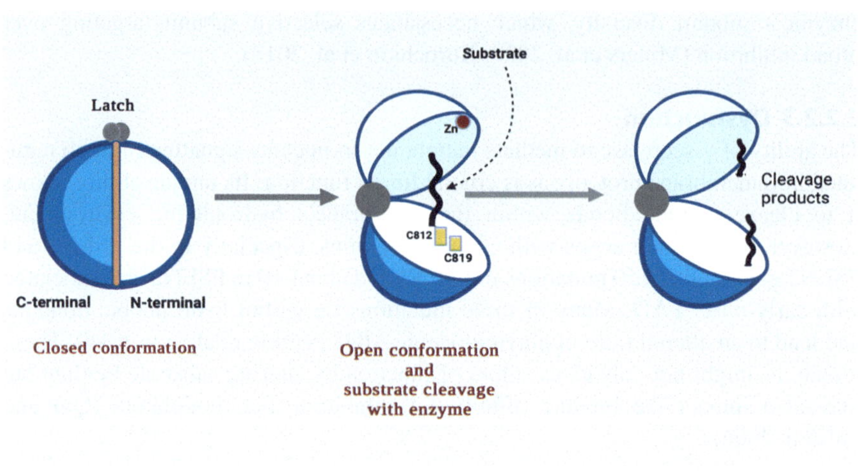

Fig. 2.6 An enzyme called insulin-degrading enzyme (IDE) degrades compounds that are connected to diabetes and Alzheimer's disease. The enclosed substrate-binding compartment of the IDE restricts substrate inflow and outflow. These conformational modifications enable IDE to accept the substrate in the open state before switching to the closed state for accurate recognition and catalysis. IDE may potentially degrade substrates that are trapped inside the catalytic chamber

2.2.3.2 Biological Role

IDE is not just a simple proteolytic enzyme; it dons several hats. Apart from its primary role in degrading a variety of substrates, IDE also functions as a chaperone/dead-end chaperone, protease modulator, E1-ubiquitin activating enzyme, and plays an integral part in the heat shock response mechanism. Its critical role in degrading Aβ peptides underscores its importance in neural health and physiology. As Aβ is secreted to the extracellular milieu via exosomes, any deviation from the proteolysis pathway activates IDE's role as a dead-end chaperone. Here, it captures and irreversibly traps Aβ monomers within its cavity, thereby curtailing the formation of harmful Aβ peptides.

2.2.3.3 Dysfunction

The clinical implications of IDE dysfunction span a spectrum of diseases. Substrates of IDE, such as insulin and amylin, have been implicated in the onset of type 2 diabetes. In contrast, the substrate Aβ has a deep-seated connection with AD. The brain, despite being replete with insulin receptors, exercises an insulin-independent uptake of glucose, highlighting insulin's multifaceted roles in neuronal functions. This intricate interplay between insulin and its receptors affects various cognitive and metabolic processes, including learning and memory. Perturbations in this balance, due to aging or disease, can have far-reaching consequences. Aging typically attenuates the levels of insulin receptors and insulin. In contrast, AD presents an anomalous upsurge in both. This, coupled with increased IDE S-nitrosylation and oxidation, impairs its proteolytic activity, laying the groundwork for Aβ aggregation and AD progression. Furthermore, a cascade of pathological events, such as Aβ and tau pathology, can be triggered by perturbations in insulin signaling (Grimmer et al. 2019; Pacheco-Quinto et al. 2013).

2.2.4 Other Relevant Enzymes

While the involvement of enzymes like gamma-secretase, BACE1, and IDE in AD is well-established, several other proteolytic entities also play pivotal roles in modulating the amyloid-beta peptide landscape. These enzymes, often lesser discussed, contribute significantly to the metabolism and clearance of amyloid-beta, influencing the progression and manifestation of the disease. Understanding their functions not only offers deeper insights into the complex biochemical dynamics of AD but also hints at potential therapeutic interventions.

2.2.4.1 NEP

A pivotal factor in the degradation of Aβ, NEP prevents its accumulation in the brain. The gene for this enzyme is situated on chromosome 3 (Palmer et al. 2010). As an integral membrane-bound metallopeptidase, NEP multitasks across a spectrum of biological processes. It functions in cell proliferation, immune response, and regulation of both cardiovascular and kidney systems. Notably, it is a neuropeptidase and an amyloid-degrading enzyme, identified in the plasma membrane as an ectoenzyme. Furthermore, NEP is a zinc-dependent enzyme, equipped with a zinc binding domain at its C-terminal. Intriguingly, it also holds the title of enkephalinase, as it is directly involved in enkephalin cleavage. However, its expression wanes with age, and such a reduction in NEP activity results in elevated oligomeric forms of Aβ. This escalation compromises hippocampal synaptic plasticity and paves the way for cognitive dysfunctions (Wang et al. 2014; Guan and Wang 2019).

2.2.4.2 Endothelin-Converting Enzyme (ECE)

Endothelin, also recognized as the endothelium converting factor, offers more than meets the eye. While endothelin 1 emerges from vascular endothelial cells, its counterparts, endothelin 2 and 3, predominantly find their place in the brain, kidney, and adrenal gland. Distinctively, these enzymes incite the inflammatory process upon a surge in Aβ levels (Talesa 2001). ECE, however, wears another hat as a critical contributor to Aβ catabolism. Its expertise lies in degrading Aβ, which effectively reduces its presence and accumulation in the brain. Classified under the elite M13 family of zinc metalloproteases, ECEs exemplify type II integral membrane-bound proteases (Musial et al. 2007; Pradeep et al. 2020).

2.2.4.3 Matrix Metalloproteinases (MMPs)

MMPs, spotlighting MMP-2 and MMP-9, have carved their niche as potential Aβ degradation agents. These zinc-fueled proteases, traditionally recognized for dismantling the extracellular matrix, inadvertently influence Aβ's behavior, dictating its aggregation and storage patterns. Their influence is far-reaching, impacting cell proliferation, migration, differentiation, apoptosis, and even angiogenesis (Wang et al. 2014).

2.2.4.4 Cyclooxygenases (COX)

Steering the inflammatory pathways are the COX enzymes, with COX-1 and COX-2 leading the charge. Their involvement in AD becomes pronounced given the pivotal role inflammation plays in its progression. By facilitating the metamorphosis of arachidonic acid into prostaglandins and their kin, they form the inflammatory bridge. An upswing in COX-2 levels in AD patients has solidified the connection between inflammation and the disease's relentless progression (Guan and Wang 2019).

2.2.4.5 Acetylcholinesterase (AChE)

AChE is a critical enzyme involved in the termination of synaptic transmission by rapidly hydrolyzing the neurotransmitter acetylcholine. Dysregulation of AChE has been observed in AD patients, leading to an accumulation of acetylcholine in the synaptic cleft and subsequently causing prolonged stimulation of postsynaptic neurons. This dysregulation and its potential as a therapeutic target make AChE inhibitors a common class of drugs used to treat AD. They work by increasing the amount and duration of acetylcholine in the brain, which can help counteract the loss of functioning synapses (Talesa 2001).

2.2.4.6 BChE

BChE, also known as pseudocholinesterase, is another cholinesterase enzyme. While its physiological function is less understood than that of AChE, it has been found to increase in the brains of AD patients. Interestingly, BChE seems to be associated with the amyloid plaques found in these brains, and its elevated levels might be compensating for decreased AChE activity. Thus, targeting both AChE and BChE can be a more effective therapeutic strategy for AD (Musial et al. 2007).

2.2.4.7 Amyloid-β (Aβ)-Degrading Enzymes

These are a group of proteolytic enzymes responsible for breaking down the amyloid-β peptide, the main component of the amyloid plaques found in AD brains. An imbalance between Aβ production and degradation is believed to be a key factor in AD onset and progression. NEP and IDE are primary members of this group (Pradeep et al. 2020).

2.2.4.8 Monoamine Oxidase B

This enzyme is involved in the catabolism of dopamine, a neurotransmitter. Inhibitors of this enzyme are used in the treatment of Parkinson's disease, another neurodegenerative disorder, but there is growing evidence that these inhibitors might also be useful in treating AD due to their neuroprotective properties (Koyiparambath et al. 2021).

2.2.4.9 Phospholipase A2

This enzyme plays a role in the release of arachidonic acid from the neuronal membranes. Given that this is a precursor for inflammatory mediators, inhibitors of this enzyme can have anti-inflammatory effects, which may benefit AD patients (Sasidharakurup et al. 2018).

2.2.4.10 Protein Kinase C

Involved in a multitude of cellular processes, dysregulation of this enzyme has implications in the abnormal processing of APP, leading to increased production of Aβ peptides (Cole et al. 1988).

2.2.4.11 11b-Hydroxysteroid Dehydrogenase Type 1

This enzyme regulates glucocorticoid action within cells. Abnormalities in its activity have been linked to chronic stress, metabolic syndrome, and AD progression.

2.2.4.12 Casein Kinase (CK-1 and CK-2)

Both these kinases have been implicated in the abnormal phosphorylation of tau protein, a major component of NFTs in AD brains. As the phosphorylation state of tau influences its ability to bind to microtubules and maintain neuronal structure, dysregulation of these kinases can exacerbate tau pathology in AD (Seckl 2010).

2.2.4.13 Cathepsin B

A lysosomal protease, Cathepsin B, is believed to play roles in both APP processing and Aβ production. Some evidence suggests that it might cleave APP in a manner that promotes Aβ production, making it another potential therapeutic target in AD (Hook et al. 2020).

Each of these enzymes provides potential avenues for therapeutic intervention in AD, and continued research into their specific roles and mechanisms might yield more effective treatments in the future.

2.3 Enzyme-Associated Models in AD

AD represents a labyrinth of interactions at the molecular level. At the epicenter are enzymes, which hold the keys to the mechanistic intricacies of this debilitating neurodegenerative disease. To grasp AD's full extent, we must delve into the roles of specific enzymes in the formation of its pathological hallmarks: amyloid-beta (Aβ) plaques and tau tangles.

2.3.1 Formation and Impact of Aβ Plaques and Tau Tangles

2.3.1.1 Aβ Plaques

Aβ is a by-product of the sequential cleavage of the APP. When Aβ peptides cluster together, they form oligomers, which then aggregate to form fibrils and eventually insoluble plaques. These plaques not only disrupt neural communication by interfering with synapses but also instigate harmful inflammatory responses, which culminate in neuron death (Ashrafian et al. 2021).

2.3.1.2 Tau Tangles

Ordinarily, the tau protein stabilizes microtubules in neurons, essential for nutrient transport in cells. However, in AD, tau becomes hyperphosphorylated, which promotes its aggregation into paired helical filaments. These filaments further assemble into NFTs, impeding neuron function and leading to their demise (Ashrafian et al. 2021). Both Aβ plaques and tau tangles are believed to contribute to the synaptic dysfunction, neuron loss, and subsequent cognitive decline observed in AD.

2.3.2 Role of Enzymes in Aβ Plaque and Tau Tangle Formation

2.3.2.1 APP Cleavage Enzymes

The generation of Aβ from APP is orchestrated by specific enzymes. BACE1 initiates the process by cleaving APP at its beta site. Subsequent cleavage by the gamma-secretase complex, which contains presenilin among its components, releases Aβ from the cell membrane (LaFerla and Oddo 2005).

2.3.2.2 Aβ Degradation and Clearance

The brain possesses mechanisms to clear Aβ, mediated by enzymes like NEP, ECE, and IDE. These enzymes break down Aβ, preventing its accumulation. With age or genetic predispositions, the efficiency of these enzymes can diminish, leading to Aβ buildup (LaFerla and Oddo 2005).

2.3.2.3 Tau Phosphorylation Regulators

The transition of tau from a functional protein to a pathological entity involves a series of phosphorylation events. Enzymes such as cyclin-dependent kinase 5 (CDK5), glycogen synthase kinase 3β (GSK3β), and casein kinases (CK-1 and CK-2) hyperphosphorylate tau. Conversely, phosphatases, like protein phosphatase 2A (PP2A), dephosphorylate tau, counteracting its aggregation. An imbalance in the activities of these enzymes can sway tau toward tangle formation (Hossain et al. 2019).

2.3.3 Other Relevant Enzymes

Several other enzymes contribute indirectly to AD's pathology like:

AChE and BChE are involved in acetylcholine breakdown, a neurotransmitter critical for memory and learning. Their elevated activity is linked to reduced acetylcholine levels in AD brains. Monoamine oxidase B participates in dopamine degradation and is linked to oxidative stress in AD. Phospholipase A2 and protein kinase C play roles in neuronal signaling and have been associated with AD-related neuroinflammation and synaptic dysfunction (Chow et al. 2010; Jakob-Roetne and Jacobsen 2009; Heneka and O'Banion 2007). By recognizing the multifaceted roles of enzymes in AD, researchers can target specific stages of Aβ and tau pathology, offering hope for more effective treatments and interventions in the future.

2.4 Strategies for Manipulation of Enzymes/Treatment

2.4.1 Current State of Research and Potential Enzyme Targets

Enzyme-based therapeutic strategies aim to target specific compounds to reestablish normal physiological metabolism, leveraging enzymes' intrinsic catalytic activities. Presently, AD treatment primarily revolves around AChE inhibitors, such as donepezil, rivastigmine, and galantamine, as well as NMDA receptor modulators like memantine. While these medications offer a temporary reprieve by enhancing cognitive functions, they address symptoms rather than the root cause. Current research highlights the paramount importance of curbing the formation of Aβ42 and mitigating the phosphorylation of the tau protein in AD treatment. Emerging studies in these domains hint at impending advancements in AD therapeutics (Yiannopoulou and Papageorgiou 2013; Lanctôt et al. 2009; Coman and Nemeş 2017).

2.4.1.1 Reduction of Aβ Production: Targeting Secretase Enzymes

β-Secretase (BACE1) Inhibitors

BACE1, an integral enzyme in the amyloidogenic pathway, processes the APP. As an aspartic protease, any inhibitor targeting it must be a hydrophilic, sizeable molecule, making crossing the BBB a challenge. Research indicates that BACE1 inhibitors, like E2609 and MK-8931, can diminish Aβ levels in human CSF by a staggering 80–90% (Godyń et al. 2016; Mathew et al. 2021).

γ-Secretase Modulation

The terminal step in the amyloidogenic processing of APP involves γ-secretase, producing Aβ40 and Aβ42 peptides. Semagacestat, a compound that appeared promising, was found to lower Aβ levels in the CSF and brain but did not halt disease progression (Steiner et al. 2018). Moreover, its administration resulted in decreased daily functioning. To sidestep the undesirable effects associated with γ-secretase inhibition, selective γ-secretase modulators (GSMs) have been proposed. GSMs tweak the enzyme's activity to modify APP processing without hindering other crucial pathways, such as the Notch signaling pathway. Some nonsteroidal anti-inflammatory drugs, including sulindac, indometacin, and flurbiprofen, show potential as GSMs (Castello and Soriano 2013).

Activation of α-Secretase

By promoting α-secretase activity, APP processing is directed more toward the non-amyloidogenic pathway, which inherently reduces the substrate availability for the amyloidogenic pathway. This shift produces a soluble Aβ peptide, acting as both a neuroprotective agent and a synaptogenesis enhancer. Epigallocatechin gallate, a flavonoid compound, has garnered interest for its dual ability to avert the formation of aberrant Aβ oligomers and stimulate α-secretase activity (Mathew et al. 2021).

2.4.1.2 Activation of Enzymes Promoting Amyloid Plaque Degradation

A cohort of proteases, including NEP, IDE, plasmin, ECE, angiotensin-converting enzyme, and metalloproteinase, can aid in the disintegration of amyloid aggregates and plaques (Steiner et al. 2018). Interestingly, the diminished activity of these enzymes in AD settings seems to correlate with the excessive buildup of amyloid plaques. Even though protease activators are a tantalizing avenue for crafting disease-altering medications, the current challenges lie in their broad specificity, which hampers their immediate clinical evaluations.

Enzymes in the Hyperphosphorylated Tau Cascade Hypothesis

Several enzymes, such as glycogen synthase kinase (GSK-3), cyclin-dependent kinase (cdk-5), microtubule affinity-regulating kinase (MARK), monoamine oxidase (MAO), and asparagine endopeptidase (AEP), have been identified and are under rigorous scrutiny for their potential roles in AD's intricate molecular tapestry (Castello and Soriano 2013).

Table 2.3 presents a detailed examination of ongoing research and potential approaches centered on enzyme manipulation as viable therapeutic targets for AD.

2.4.2 Designing Enzyme Inhibitors: Challenges and Potential Approaches

The amyloid β peptide (Aβ), found in the senile plaques of AD patients, arises from the proteolytic processing of the APP. BACE1 is integral to Aβ formation as it cleaves the N-terminus of the Ab domain (Payne et al. 2022). Notably, the BACE1 inhibitors, KMI-358 and KMI-370, have demonstrated efficacy. The substitution of the Glu residue at the P4 position with a β-*N*-oxalyl-DAP (L-diamino propionyl) residue is reported to augment their inhibitory action against BACE1.

A current trend in drug design is to formulate compounds that concurrently target multiple enzymes. The structural scaffolds of AChEI drugs like donepezil and tacrine are prime examples of this approach, well-documented for their efficacy (O'Brien and Wong 2011). An innovative method merges the pharmacophores of AChEI drugs and BACE-1 inhibitors using a specific linker, yielding a dual inhibitor targeting both enzymes (Zhang et al. 2019).

Investigations assessing the inhibitory activity of compounds with varying carbon atom numbers between benzene and piperidine revealed a stark decline in inhibitory prowess against cholinesterases (ChEs) for those with only two carbon atoms. Such diminished activity may be attributed to the substantial molecular volume of these compounds, obstructing their access to the active pockets of ChEs. The

Table 2.3 The current state of research and potential strategies for enzyme manipulation as therapeutic targets for AD

S. no.	Enzyme target	Drug	Potential therapeutic strategy	Clinical phases	Current state of research challenges/ limitations or ADR	References
1.	α-Secretase	LY450139 (Semagacestat)	Inhibitor	Phase 3	Fatigue, lethargy, asthenia, diarrhea, increase in mean Fridericia QTc	Zhang et al. (2019)
2.	β-Secretase 1 (BACE1)	CNP520 (Umibecestat)	Inhibitor	Phase 2	Prolongation of QT interval and Torsade's de pointes, lethal arrhythmia	Henley et al. (2009)
3.	Acetylcholinesterase	Huperazine-A	Inhibitor	Phase 2	Tachycardia, low energy, dry mouth, hypertension, muscle cramps, arthralgia, nausea, drowsiness, and diarrhea	Vormfelde et al. (2020)
4.	Acetylcholinesterase	Ladostigil	Inhibitor	Phase 2	Atrial fibrillation, depression, prostatic hypertrophy, extremity pain	Little et al. (2008)
5.	Monoamine oxidase B	Rasagaline	Inhibitor	Phase 2	Tendon rupture, agitation/irritability, or psychosis	Schneider et al. (2019), Matthews et al. (2021)
6.	Phospholipase A2	Rilapladib	Inhibitor	Phase 2	Headache, dizziness, nausea, urinary tract infection, diarrhea, cystitis, fatigue	Lefaucheur et al. (2014)
7.	Protein kinase C	Bryostatin 1	Activator	Phase 2	Myalgia, fatigue, lethargy, low-grade pyrexia, nausea, and anorexia	Maher-Edwards et al. (2015)
8.	11β-hydroxysteroid dehydrogenase type 1	Xanamem	Inhibitor	Phase 2	Headache, thrombophlebitis, diarrhea	Sun and Alkon (2006)
9	β-Secretase	E2609	Inhibitor	Phase 1	Not provided	Webster et al. (2017)
10	α-Secretase	Epigallocatechin gallate	Enhancer/amyloid aggregation inhibitor	Phase 2/3	Not provided	Lynch et al. (2018)

N-benzylpiperidine residue emerges as pivotal for inhibiting both ChEs (Zhu et al. 2009).

Compounds containing a piperazine ring, as opposed to a piperidine ring, may exhibit comparable or enhanced inhibitory action on ChEs. Using the structure of donepezil as a template, several researchers have conjugated benzylpiperidine/benzylpiperazine with biologically active heterocyclic derivatives. These compounds exhibited promising inhibitory activity against AChE, antioxidant properties, Aβ peptide aggregation inhibition, and metal chelation (Guo et al. 2020).

2.4.2.1 AChE and BACE-1 Multi-target Strategy

Research on a compound merging donepezil with 2,4-disubstituted pyrimidine demonstrated dual enzyme inhibition, with marked inhibitory activities against AChE and BACE-1, and further displayed reduced Aβ1-40 aggregation and enhanced neuroblastoma cell viability (Mohamed et al. 2011). In another study, the Munoz–Torrero team synthesized an AChEI hybrid by fusing tacrine and 4-oxo-4H-chromene, revealing potent inhibitory actions against both AChE and BACE-1, coupled with a significant decrease in Aβ42 aggregation (Pérez-Areales et al. 2017). The Bartolini group's contribution was a robust hybrid compound that married tacrine with benzofuran. This fusion not only targeted AChE and BACE-1 but also significantly reduced self-induced Aβ aggregation. Additionally, the hybrid showcased improved cognitive function in scopolamine-treated ICR mice without evident hepatotoxicity (Zha et al. 2016). A subsequent study led by Belluti introduced a benzophenone-core compound, which effectively inhibited both AChE and BACE-1 (Belluti et al. 2014). In a recent breakthrough, the Gabr team reported a compound, also with a benzophenone core. Apart from its profound inhibition of both AChE and BACE-1, this compound exhibited metal chelation properties, limited the toxicity to SH-SY5Y neuroblastoma cells, and proved its ability to penetrate the BBB through PAMPA evaluation (Gabr and Abdel-Raziq 2018).

2.4.2.2 AChE and GSK-3β Multi-target Strategy

In one notable study by Hui et al., a hybrid substance comprising the AChEI tacrine and phenothiazine (pertinent for tau aggregation) exhibited significant inhibitory potential against AChE. This hybrid was also efficacious in hindering Aβ self-aggregation (Hui et al. 2014).

2.4.2.3 BACE-1 and GSK-3β Multi-target Strategy

In AD, senile plaques and NFTs are predominant brain markers. BACE-1 and GSK-3β are two vital targets behind these markers. Simultaneously inhibiting both could present synergistic benefits (Prati et al. 2018). The Cavalli group created a series of triazinone derivatives as dual inhibitors of BACE-1 and GSK-3β. One particular compound displayed balanced inhibitory potential against both targets and lacked neurotoxicity in glial and neuronal cells. This compound's effectiveness is attributed to its ability to modulate GSK-3β activity and inhibit both targets, leading to both neuroprotective and neurogenic effects (Prati et al. 2015). Some 2-furoyl piperazine-based sulfonamide derivatives have emerged as promising compounds for AD (Hassan et al. 2019).

2.4.2.4 AChE and MAOs Multi-target Strategy

Ladostigil, a prospective treatment for Alzheimer's type dementia, is presently under Phase II clinical trials (Weinreb et al. 2016). This drug showcases inhibitory effects against AChE, BChE, MAO-A, and MAO-B in the brain (Rosini et al. 2016). Its molecular design combines the carbamate moiety of rivastigmine, a ChE inhibitor, with the indolamine segment of rasagiline, an MAO-B inhibitor.

Using AChE and MAOs targets for AD treatments has gained popularity with donepezil and tacrine-related derivatives serving as multi-target directed ligands. A standout compound fuses the benzylpiperidine segment of donepezil with the indolyl propargyl amino segment of an MAOs inhibitor. This compound demonstrated multi-target inhibitory power against AChE, BChE, MAO-A, and MAO-B and showcased considerable anti-aggregation properties (Bolea et al. 2011).

2.5 Regulation and Impacts of Enzymes

2.5.1 Regulation of Enzymes with Age

As aging occurs, BACE activity in the brains of mice, monkeys, and humans intensifies significantly across all brain regions. However, BACE protein levels remain constant. This suggests that the age-related surge in BACE activity might be a key player behind the Aβ production and accumulation in the brain, elevating the risk of AD in humans (Fukumoto et al. 2004). Research hints that aging might influence γ-secretase activity and specificity, playing a role in sporadic AD. An elevated γ-secretase activity in aged females might be linked to the higher sporadic AD incidence in women and the more aggressive Aβ plaque pathology observed in female AD mouse models. Decreased Notch γ-secretase activity could also play a part in neurodegeneration (Placanica et al. 2009).

Enzymes like NEP, NEP2, ECEs (ECE-1 and ECE-2), and IDE are principal players in degrading brain Aβ. Aging and certain pathologies like ischemia and stroke reduce their expression and activity, leading to diminished Aβ clearance and subsequent accumulation in the brain. Research on AD induced in rats through the APP/presenilin-1 (PS-1) showed NEP protein and activity decreased significantly with age, especially in the cerebral cortex and hippocampus of APP/PS1 mice over 6 months of age. Meanwhile, ECE protein increased with age, but its activity lessened in the same regions of older APP/PS1 mice (Apelt et al. 2003). Aging is also correlated with elevated AEP levels and an increase in fragmented APP levels (Zhang et al. 2016).

2.5.2 Other Conditions Causing Elevation of Enzymes

Investigations on rat retina demonstrated that mitochondrial respiration obstruction and oxidative stress can lead to an upsurge in β-secretase (BACE1) proteins and activity (Xiong et al. 2007). A noteworthy elevation in AChE activity was discerned in the islets of Langerhans, surpassing that in exocrine tissues of the rat pancreas. Furthermore, rats suffering from streptozotocin-induced diabetes exhibited higher levels compared to their healthy counterparts (Godfrey and Matschinsky 1975). Amplified concentrations of both AChE and butyrylcholinesterase can induce a decline in acetylcholine levels, potentially spiking the levels of IL-1 and TNF-a due to the disruption of the Ach-mediated feedback loop (Rao et al. 2007). There's a noteworthy connection between mutations in γ-secretase substrates and several human ailments, spanning from cancer to vascular and skin dysfunctions (Jurisch-Yaksi et al. 2013). Individuals with heart failure presented elevated concentrations of soluble NEP, a significant predictor of cardiovascular mortality and related hospitalizations (Bayés-Genís et al. 2015). A study indicated that the asparagine endopeptidase inhibitor aids in shielding against fenpropathrin-induced neurodegeneration, mitigating α-synuclein aggregation and neuroinflammation, hinting at its relevance in Parkinsonism (Yu et al. 2020). The IDE, pivotal in dismantling peptides like insulin and amyloid (A), plays a quintessential role in thwarting toxic compound formation and ensuring proteostasis. A deficit in IDE is intertwined with the onset of type 2 diabetes (Farris et al. 2003).

2.5.3 Regulators of AD

2.5.3.1 BACE Regulators

Various factors can upregulate BACE-1 mRNA, both in vitro and in vivo. These include hypoxia and oxidative stress. The hypoxia-inducible factor-1α has shown affinity to bind to the BACE-1 promoter. Meanwhile, oxidative stress can induce BACE-1 expression via pathways such as c-jun N-terminal kinase and stress-activated protein kinase (Tamagno et al. 2008). Cyclin-dependent kinase 5 activation is another trigger for increased levels of BACE-1 mRNA and its protein (Wen et al. 2008). Traumatic brain injury and infection by herpes simplex virus1 (HSV-1) can also enhance BACE-1 expression (Wozniak et al. 2007). Notably, HSV-1, particularly when associated with the ε4 allele of the apolipoprotein E4 gene, is a recognized risk factor for AD. Viral DNA has even been detected within amyloid plaques in AD-afflicted brains (Lammich et al. 2004).

Recent studies indicate that cellular energy deprivation, or glucose deprivation in cells, results in a post-transcriptional increase in BACE-1 levels. This rise can be attributed to enhanced eIF2α phosphorylation, a finding corroborated in AD transgenic mice (Tg2576). Chronic energy inhibition in these mice showed increased eIF2α phosphorylation, elevated BACE-1 levels, and enhanced amyloidogenesis (Tesco et al. 2007). In situations of stress, like hypoxia/ischemia or viral infection, BACE-1 levels can be influenced through eIF2α-regulated translation initiation.

Additionally, the lysosomal and proteasomal pathways, coupled with lysine acetylation status, can further dictate BACE-1 protein stability (Kojro et al. 2010).

2.5.3.2 Alpha Secretase Regulators

Alpha secretase cleavage is influenced by various molecules. Some inhibitors include tissue inhibitor of metalloproteases 3 (Postina 2012), and alterations in membrane cholesterol concentration also play a role. Alpha secretase cleavage generally happens outside cholesterol-rich membrane microdomains or lipid rafts. Inhibiting cholesterol synthesis, via agents like statins or the squalene synthase inhibitor zaragozic acid, promotes alpha secretase cleavage (Zhou et al. 2007). Moreover, compounds like phorbol esters, protein kinase A and C (PKC) activators, muscarinic agonists, neuropeptides, statins, and retinoids have been identified to augment alpha-secretase processing of APP (Pivovarova et al. 2016).

2.5.3.3 Gamma Secretase Regulators

CD147, a membrane protein, is recognized alongside various proteins like GSK-3, PLD1, and TMP21 as gamma secretase regulators. Moreover, factors like cholesterol metabolism, the lipid bilayer environment, and inflammatory cytokines such as TNF-R, IFN-γ, and IL-1β stimulate γ-secretase activity. These elements also exhibit a physical and functional association with γ-secretase complexes (Belyaev et al. 2009).

2.5.3.4 IDE Regulators

The activity of IDE can be modulated by various compounds. Compounds such as apomorphine, somatostatin, curcumin, docosahexaenoic acid, leptin, propranolol, and suramin have been found to increase IDE activity. In contrast, compounds like bacitracin, sevoflurane, adenosine triphosphate, and palmitic acid have been reported to decrease its activity (Nalivaeva et al. 2012).

2.5.3.5 NEP Regulators

Recent research highlights the potential of histone deacetylase (HDAC) inhibitors in inducing NEP expression. For example, valproic acid has been shown to amplify NEP activity and expression in SHSY-5Y cells and in specific regions of rats such as the cortex and hippocampus (Eisele et al. 2007). Valproic acid has also been found beneficial in reversing cognitive deficits induced by hypoxia in rats. Similarly, trichostatin A, another HDAC inhibitor, enhanced NEP expression in SHSY-5Y cells (Klein et al. 2018). Imatinib (Gleevec) is observed to elevate NEP's mRNA and protein levels in various cell lines (Oefner et al. 2000). Recent findings also suggest 5-hydroxyindoleacetic acid (5-HIAA) as a potential inducer of NEP levels, both in neuroblastoma cells and in vivo (Oefner et al. 2007).

However, some agents, such as phosphoramidon and tiorphan, have been recognized as NEP inhibitors (Campbell 2017). These agents obstruct the NEP active site, preventing substrate binding. MCB3937 has also been identified as an NEP inhibitor with a potent IC50 value in the nanomolar range (Prinzen et al. 2005). LCZ696, combining structural elements of valsartan and AHU377, is another noteworthy NEP inhibitor. Notably, AHU377 metabolizes into LBQ657, an established NEP inhibitor, with IC50 values of 2.4 and 2.3 nM for valsartan and LBQ657 respectively against NEP (Kim and Lee 2020).

2.5.3.6 Regulators of ADAM10

ADAM10 protein levels can be elevated by compounds like retinoic acid and its derivative, acitretin. Studies on SH-SY5Y cells, when exposed to 1 μM of retinoic acid over 4 days, revealed an upsurge in ADAM10 mRNA levels (Vassar et al. 2009). Compounds such as Am80, bryostatin-1, and phlogacantholide C have also been identified to enhance ADAM10 levels, promoting the cleavage of APP and increasing the secretion of ADAM10-induced APP fragments.

Resveratrol, a polyphenol present in foods like grape skin and pomegranates, has been observed to indirectly boost ADAM10 expression. Mice experiments indicate that administering resveratrol for 4–5 days results in reduced plaque formation. Furthermore, both gemfibrozil and etazolate have demonstrated potential in raising ADAM10 levels. In studies on rat cortical neurons, etazolate was found to stimulate ADAM10 in a concentration-dependent manner within the range of 0.2–20 μM (Lee et al. 2020).

2.5.4 Substrates

2.5.4.1 BACE Substrates

Beyond APP, BACE1 interacts with a plethora of substrates. Recognizing these substrates not only provides insights into potential toxicity that could arise from BACE1 inhibition but also aids in the design of precise and efficient BACE1 inhibitors. The majority of BACE1 substrates are transmembrane proteins. Notable among these are the APP and its like-proteins, APLP1 and APLP2. These molecules play pivotal roles in modulating neurotransmission and synaptic plasticity within the central nervous system. Nevertheless, their actions can sometimes lead to undesired outcomes, including disrupted cell–cell signaling, hindered injury recovery, impaired synaptic functions, and memory lapses. Other substrates, such as P-selectin glycoprotein ligand-1, the low-density lipoprotein receptor-related protein, neuregulin-1, and neuregulin-3, are vital for synapse upkeep. However, they too can exhibit harmful effects, such as compromised remyelination following nerve damage (Kopan and Ilagan 2004).

2.5.4.2 γ-Secretase Substrates

The list of γ-secretase substrates and potential candidates surpasses 140, all of which are type 1 transmembrane proteins. A few familiar names include APLP1, APLP2, and APP, all of which regulate synaptic plasticity and neuronal excitability. A slew of other substrates play roles in cell adhesion (e.g., N-cadherin, E-cadherin, CD44), protein tyrosine kinase (like the CSF1 receptor), and growth factor-dependent receptor tyrosine kinase (ErbB4). Moreover, signaling receptors such as Notch 1–4, Notch ligands (Delta and Jagged), a neurotrophin co-receptor (p75), and a cell-surface proteoglycan co-receptor (syndecan-3) are also on the list (Lichtenthaler 2011). Intriguingly, the sequence of a substrate does not influence the cleavage by γ-secretase. Instead, this process is determined by ectodomain shedding. In numerous instances, the ICDs liberated following γ-secretase cleavage participate in gene transcription regulation. For instance, Notch, a renowned substrate, first undergoes ectodomain shedding at the S2 site through metalloprotease actions. Subsequently, γ-secretase cleaves it at the S3 site, which leads to the release of the Notch intracellular domain (NICD) (Sousa et al. 2021).

2.5.4.3 α-Secretase and IDE

α-Secretase activity is chiefly associated with ADAM10 and ADAM17, both integral components of the ADAMs family of type I integral membrane proteins. Even though they hail from the same protein family, ADAM10 and ADAM17 demonstrate varied substrate targets and functional modes. Among their most notable substrates are pro-TNF, collagen XVII, and desmoglin 2. These substrates have critical functions in regulating inflammation, cell motility, and cell adhesion, respectively (Yu et al. 2016).

Separately, IDE is not solely devoted to insulin. It recognizes and acts upon a range of substrates. Notably, it targets glucagon, insulin-like growth factors (IGF-I and IGF-II), amyloid-β peptides (A-β), and the islet amyloid polypeptide (IAPP), which is commonly referred to as amylin. Each of these substrates plays diverse roles in the body's metabolism and cellular functions (Jonsson et al. 2013).

2.6 Future Direction

Neuropsychology has played a pivotal role in unveiling the cognitive changes associated with AD and other related dementias. Our enhanced understanding of AD, coupled with our ability to distinguish it from other conditions, has shed light on the cognitive changes manifesting during disease progression and its correlation with aging. Cutting-edge developments have now made possible the inception of biomarkers, capable of discerning the neuropathologic alterations within individuals and detecting the earliest transformations. Future endeavors in precision medicine promise to curate specialized treatment methodologies tailored for individuals, accentuating the importance of early diagnosis when therapeutic interventions might be at their most potent (Bondi et al. 2017).

Treatment of AD is, however, a multifaceted challenge. Its origins range from protein irregularities and inflammation to nerve damage and insulin deregulation. Repurposing existing drugs presents a beacon of hope. Some of the most promising candidates in this arena encompass anti-inflammatory, neuroprotective, and antidiabetic drugs. As our comprehension of protein biology—particularly Aβ and tau—evolves, our therapeutic strategies for AD will refine. This profound understanding could pave the way for innovative compounds and immunotherapies, potentially revolutionizing patient care (Yu et al. 2021).

The current treatment paradigms for AD primarily center around three strategies, with anti-amyloid therapy being the most prominent. This encompasses secretase inhibitors, Aβ aggregation inhibitors, and Aβ immunotherapy. Novel approaches, such as RIPK1, GSK-3β, Fyn kinase inhibition, and CB2 receptor agonists, are also being explored. While the anti-Aβ immunotherapy avenue, particularly passive immunotherapy employing exogenous mAbs, has been the focal point of extensive research, not all clinical trials have been fruitful. The setbacks encountered with BACE1 inhibitors, especially in phase III clinical trials, have urged researchers to seek therapeutic agents divergent from the prevailing amyloid hypothesis. Despite the longstanding research on secretase inhibitors, their true efficacy remains a topic of contention (Yu et al. 2021; Moussa-Pacha et al. 2020).

The natural realm too holds promise in the battle against AD. Certain compounds naturally inhibit Aβ aggregation, a prospective strategy in AD prevention. Endeavors to target brain chaperones and metals, inhibiting the interaction between Aβ peptides and these metals, have the potential to halt Aβ oligomerization. Modern biophysical and structural biology techniques have identified promising compounds like tanshinone and uncarinic acid C, known for their binding epitopes or planar hydrophobic structures. Additionally, compounds such as epigallocatechin gallate (EGCG) and resveratrol have demonstrated an affinity toward the toxicity determinants of Aβ, specifically targeting the N-terminus and β1-turn regions (Menard et al. 2013).

The path ahead is riddled with challenges, but the concerted efforts of the scientific community provide a beacon of hope for the future. Several compounds, notably EGCG, oleuropein aglicone (OleA), and quercetin, have emerged as potential therapeutic agents for AD (Henríquez et al. 2020).

To alleviate the Aβ burden, both active and passive immunotherapy methods are in the fray. Noteworthy monoclonal antibodies that have progressed to phase III trials include gantenerumab, aducanumab, and BAN2401. These antibodies have demonstrated their capacity to target amyloid oligomers, subsequently reducing tau levels. Specifically, BAN2401 exhibited cognitive and functional benefits in patients with MCI to mild AD, especially those bearing the APOE4 gene. Furthermore, drugs targeting Aβ oligomers like CT1812, PQ912, and ALZ-801 are in the pipeline. Concurrently, for anti-tau therapy, avenues under exploration include phosphatase modifiers, kinase inhibitors, tau aggregation inhibitors, microtubule stabilizers, and tau immunotherapy. The progression and severity of AD bear a significant relationship to neuroinflammation. Emerging PET targets, capable of discerning neuroinflammation, present new therapeutic strategies. These targets, encompassing the likes of cannabinoid receptor 2, COX1 or COX2, P2X7 receptor, lysosomal proteins, and astrocytes, are intricately associated with amyloid plaques (Calsolaro and Edison 2016). However, challenges persist in certain interventions, demanding standardization for efficacy. The repetitive transcranial magnetic stimulation technique exemplifies this predicament. Given the variance in efficacy across AD progression stages, the future might necessitate multiple standardized protocols contingent on disease severity. Advancements in retinal imaging technology present a paradigm shift in AD research. Their cost-effectiveness, noninvasiveness, and clinical adaptability make them promising. While preliminary studies hint at a correlation between retinal measures and AD, comprehensive replication and longitudinal research are imperative. A standardized framework for retinal imaging in AD, coupled with artificial intelligence, can potentially revolutionize early risk assessment and stratification (Cheung et al. 2021).

The advent of induced pluripotent stem cells from human somatic cells has ushered in novel approaches to understanding AD and devising therapeutic interventions. Animal model-based studies underscore the potential of cell replacement therapies in curbing AD pathology and bolstering cognitive function (Wang et al. 2022).

Personalized treatment mandates consideration of diverse patient facets—age, gender, genetics, medical history, environmental factors, and lifestyle. This holds true across pharmacological and non-pharmacological interventions. The strides in machine learning enable us to craft models predicting optimal therapeutic strategies for AD, accommodating the multifaceted nature of the disease. Further research, aiming to elucidate the interplay between patient attributes and therapeutic responses, is paramount. This continued exploration could pave the way for precision medicine, heralding a new era of hope for AD patients.

2.7 Conclusion

AD, a formidable global health concern, is characterized by gradual cognitive decline and memory loss. Central to its pathogenesis is the accumulation of extracellular amyloid-beta (Aβ) plaques and intracellular NFTs, largely attributed to the enzymatic processes mediated by beta-secretase 1 (BACE1), gamma-secretase, NEP, and IDE, among others. These enzymes play pivotal roles in the synthesis and degradation of Aβ peptides, influencing their aggregation and clearance.

While genetic factors, like mutations in the APP, PSEN1, or PSEN2 genes, contribute to familial AD, the ε4 variant of the APOE gene stands as a significant risk factor for the more prevalent late-onset AD. Recognizing the intricacies of these enzymes offers a lens through which the progression of AD can be understood and potentially manipulated. Current studies have accentuated the potential of enzyme modulation—whether through inhibition, enhancement, or novel therapeutic strategies—as a promising avenue for AD treatment. Notably, breakthroughs in medicinal chemistry based on enzyme structures have kindled hope for small molecule inhibitors that can traverse the BBB.

Furthermore, enzymes like AChE and monoamine oxidase B, which are involved in neurotransmitter breakdown and linked to oxidative stress, respectively, offer additional targets for intervention. By delving deeper into the multifaceted roles of these enzymes in AD, researchers stand at the cusp of not only devising more effective treatments but also unraveling the intricate etiology of this devastating disease. The collective knowledge underscores the paramount importance of enzymes in AD progression and offers a beacon of hope for future therapeutic advancements.

References

Ahmad SS, Khan S, Kamal MA, Wasi U (2019) The structure and function of α, β and γ-secretase as therapeutic target enzymes in the development of Alzheimer's disease: a review. CNS Neurol Disord Drug Targets 18(9):657–667

Alam J, Sharma L (2018) Potential enzymatic targets in Alzheimer's: a comprehensive review. Curr Drug Targets 20(3):316–339

Alam J, Sharma L (2019) Potential enzymatic targets in Alzheimer's: a comprehensive review. Curr Drug Targets 20(3):316–339. https://pubmed.ncbi.nlm.nih.gov/30124150/

Altuna-Azkargorta M, Mendioroz-Iriarte M (2018) Blood biomarkers in Alzheimer's disease. Neurologia 36(9):704–710. http://www.ncbi.nlm.nih.gov/pubmed/29752036

Apelt J, Ach K, Schliebs R (2003) Aging-related down-regulation of neprilysin, a putative β-amyloid-degrading enzyme, in transgenic Tg2576 Alzheimer-like mouse brain is accompanied by an astroglial upregulation in the vicinity of β-amyloid plaques. Neurosci Lett 339(3):183–186

Ashrafian H, Zadeh EH, Khan RH (2021) Review on Alzheimer's disease: inhibition of amyloid beta and tau tangle formation. Int J Biol Macromol 167:382–394

Bayés-Genís A, Barallat J, Galán A, De Antonio M, Domingo M, Zamora E et al (2015) Soluble neprilysin is predictive of cardiovascular death and heart failure hospitalization in heart failure patients. J Am Coll Cardiol 65(7):657–665

Bazzari FH, Bazzari AH (2022) BACE1 inhibitors for Alzheimer's disease: the past, present and any future? Molecules 27(24):8823

Belluti F, De Simone A, Tarozzi A, Bartolini M, Djemil A, Bisi A et al (2014) Fluorinated benzophenone derivatives: balanced multipotent agents for Alzheimer's disease. Eur J Med Chem 78:157–166. https://pubmed.ncbi.nlm.nih.gov/24681980/

Belyaev ND, Nalivaeva NN, Makova NZ, Turner AJ (2009) Neprilysin gene expression requires binding of the amyloid precursor protein intracellular domain to its promoter: implications for Alzheimer disease. EMBO Rep 10(1):94–100

Bolea I, Juárez-Jiménez J, De Los RC, Chioua M, Pouplana R, Luque FJ et al (2011) Synthesis, biological evaluation, and molecular modeling of donepezil and N-[(5-(benzyloxy)-1-methyl-1H-indol-2-yl)methyl]-N-methylprop-2-yn-1-amine hybrids as new multipotent cholinesterase/monoamine oxidase inhibitors for the treatment of Alzheimer's disease. J Med Chem 54(24):8251–8270

Bondi MW, Edmonds EC, Salmon DP (2017) Alzheimer's disease: past, present, and future. J Int Neuropsychol Soc 23(9–10):818–831. https://pubmed.ncbi.nlm.nih.gov/29198280/72

Breijyeh Z, Karaman R (2020) Comprehensive review on Alzheimer's disease: causes and treatment. Molecules 25:5789

Calsolaro V, Edison P (2016) Neuroinflammation in Alzheimer's disease: current evidence and future directions. Alzheimers Dement 12(6):719–732

Campbell DJ (2017) Long-term neprilysin inhibition-implications for ARNIs. Nat Rev Cardiol 14:171–186

Castello MA, Soriano S (2013) Rational heterodoxy: cholesterol reformation of the amyloid doctrine. Ageing Res Rev 12:282–288

Cervellati C, Trentini A, Rosta V, Passaro A, Bosi C, Sanz JM et al (2020) Serum beta-secretase 1 (BACE1) activity as candidate biomarker for late-onset Alzheimer's disease. Geroscience 42(1):159–167

Chen S, Mima D, Jin H, Dan Q, Wang F, Cai J et al (2021) The association between neprilysin gene polymorphisms and Alzheimer's disease in Tibetan population. Brain Behav 11(3):e02002. https://pubmed.ncbi.nlm.nih.gov/33314757/

Cheung CY, Mok V, Foster PJ, Trucco E, Chen C, Wong TY (2021) Retinal imaging in Alzheimer's disease. J Neurol Neurosurg Psychiatry 92(9):983–994. https://pubmed.ncbi.nlm.nih.gov/34108266/

Chow VW, Mattson MP, Wong PC, Gleichmann M (2010) An overview of APP processing enzymes and products. NeuroMolecular Med 12:1–2

Cole S, Vassar R (2008) BACE1 structure and function in health and Alzheimer's disease. Curr Alzheimer Res 5(2):100–120. https://pubmed.ncbi.nlm.nih.gov/18393796/

Cole G, Dobkins KR, Hansen LA, Terry RD, Saitoh T (1988) Decreased levels of protein kinase C in Alzheimer brain. Brain Res 452(1–2):165–174

Coman H, Nemeş B (2017) New therapeutic targets in Alzheimer's disease. Int J Gerontol 11:2–6

De Strooper B, Vassar R, Golde T (2010) The secretases: enzymes with therapeutic potential in Alzheimer disease. Nat Rev Neurol 6:99–107

Eisele YS, Baumann M, Klebl B, Nordhammer C, Jucker M, Kilger E (2007) Gleevec increases levels of the amyloid precursor protein intracellular domain and of the amyloid-degrading enzyme neprilysin. Mol Biol Cell 18:3591–3600. http://www.molbiolcell.org/cgi/doi/10.1091/mbc.E07-01-0035

Farris W, Mansourian S, Chang Y, Lindsley L, Eckman EA, Frosch MP et al (2003) Insulin-degrading enzyme regulates the levels of insulin, amyloid-protein, and the-amyloid precursor protein intracellular domain in vivo. Proc Natl Acad Sci U S A 100:4162

Fukumoto H, Rosene DL, Moss MB, Raju S, Hyman BT, Irizarry MC (2004) β-Secretase activity increases with aging in human, monkey, and mouse brain. Am J Pathol 164(2):719–725

Gabr MT, Abdel-Raziq MS (2018) Design and synthesis of donepezil analogues as dual AChE and BACE-1 inhibitors. Bioorg Chem 80:245–252

Godfrey DA, Matschinsky FM (1975) Enzymes of the cholinergic system in islets of Langerhans. J Histochem Cytochem 23:645

Godyń J, Jończyk J, Panek D, Malawska B (2016) Therapeutic strategies for Alzheimer's disease in clinical trials. Pharmacol Rep 68(1):127

Grimmer T, Goldhardt O, Yakushev I, Ortner M, Sorg C, Diehl-Schmid J et al (2019) Associations of neprilysin activity in CSF with biomarkers for Alzheimer's disease. Neurodegener Dis 19(1):43–50. https://pubmed.ncbi.nlm.nih.gov/31266021/

Guan PP, Wang P (2019) Integrated communications between cyclooxygenase-2 and Alzheimer's disease. FASEB J 33(1):13–33

Guo Y, Yang H, Huang Z, Tian S, Li Q, Du C et al (2020) Design, synthesis, and evaluation of acetylcholinesterase and butyrylcholinesterase dual-target inhibitors against Alzheimer's diseases. Molecules 25(3):489

Hassan M, Abbasi MA, Aziz-ur-Rehman, Siddiqui SZ, Shahzadi S, Raza H et al (2019) Designing of promising medicinal scaffolds for Alzheimer's disease through enzyme inhibition, lead optimization, molecular docking and dynamic simulation approaches. Bioorg Chem 91:103138

Heneka MT, O'Banion MK (2007) Inflammatory processes in Alzheimer's disease. J Neuroimmunol 184(1–2):69–91

Henley DB, May PC, Dean RA, Siemers ER (2009) Development of semagacestat (LY450139), a functional gamma-secretase inhibitor, for the treatment of Alzheimer's disease. Expert Opin Pharmacother 10(10):1657–1664. https://pubmed.ncbi.nlm.nih.gov/19527190/

Henríquez G, Gomez A, Guerrero E, Narayan M (2020) Potential role of natural polyphenols against protein aggregation toxicity: in vitro, in vivo, and clinical studies. ACS Chem Neurosci 11(19):2915–2934

Hook V, Yoon M, Mosier C, Ito G, Podvin S, Head BP, Rissman R, O'Donoghue AJ, Hook G (2020) Cathepsin B in neurodegeneration of Alzheimer's disease, traumatic brain injury, and related brain disorders. Biochim Biophys Acta Proteins Proteom 1868(8):140428

Hossain MF, Uddin MS, Uddin GS, Sumsuzzman DM, Islam MS, Barreto GE, Mathew B, Ashraf GM (2019) Melatonin in Alzheimer's disease: a latent endogenous regulator of neurogenesis to mitigate Alzheimer's neuropathology. Mol Neurobiol 56:8255–8276

Hui AL, Chen Y, Zhu SJ, Gan CS, Pan J, Zhou A (2014) Design and synthesis of tacrine-phenothiazine hybrids as multitarget drugs for Alzheimer's disease. Med Chem Res 23(7):3546–3557

Hur JY (2022) γ-Secretase in Alzheimer's disease. Exp Mol Med 54(4):433–446. https://pubmed.ncbi.nlm.nih.gov/35396575/

Jakob-Roetne R, Jacobsen H (2009) Alzheimer's disease: from pathology to therapeutic approaches. Angew Chem Int Ed 48(17):3030–3059

Jonsson T, Stefansson H, Steinberg S, Jonsdottir I, Jonsson PV, Snaedal J et al (2013) Variant of TREM2 associated with the risk of Alzheimer's disease. N Engl J Med 368(2):107–116

Jurisch-Yaksi N, Sannerud R, Annaert W (2013) A fast growing spectrum of biological functions of γ-secretase in development and disease. Biochim Biophys Acta Biomembr 1828:2815–2827

Kim N, Lee HJ (2020) Target enzymes considered for the treatment of Alzheimer's disease and Parkinson's disease. Biomed Res Int 2020:2010728

Klein C, Roussel G, Brun S, Rusu C, Patte-Mensah C, Maitre M et al (2018) 5-HIAA induces neprilysin to ameliorate pathophysiology and symptoms in a mouse model for Alzheimer's disease. Acta Neuropathol Commun 6(1):136

Kojro E, Füger P, Prinzen C, Kanarek AM, Rat D, Endres K et al (2010) Statins and the squalene synthase inhibitor zaragozic acid stimulate the non-amyloidogenic pathway of amyloid-β protein precursor processing by suppression of cholesterol synthesis. J Alzheimers Dis 20(4):1215–1231

Kopan R, Ilagan MXG (2004) γ-Secretase: proteasome of the membrane? Exp Mol Med 5(6):499–504

Koyiparambath VP, Prayaga Rajappan K, Rangarajan TM, Al-Sehemi AG, Pannipara M, Bhaskar V, Nair AS, Sudevan ST, Kumar S, Mathew B (2021) Deciphering the detailed structure–activity relationship of coumarins as monoamine oxidase enzyme inhibitors—an updated review. Chem Biol Drug Des 98(4):655–673

Kurochkin IV, Guarnera E, Berezovsky IN (2018) Insulin-degrading enzyme in the fight against Alzheimer's disease. Trends Pharmacol Sci 39(1):49–58. https://pubmed.ncbi.nlm.nih.gov/29132916/

LaFerla FM, Oddo S (2005) Alzheimer's disease: Aβ, tau and synaptic dysfunction. Trends Mol Med 11(4):170–176

Lammich S, Schöbel S, Zimmer AK, Lichtenthaler SF, Haass C (2004) Expression of the Alzheimer protease BACE1 is suppressed via its 5′-untranslated region. EMBO Rep 5(6):620–625

Lanctôt KL, Rajaram RD, Herrmann N (2009) Therapy for Alzheimer's disease: how effective are current treatments? Ther Adv Neurol Disord 2:163–180

Lauer AA, Mett J, Janitschke D, Thiel A, Stahlmann CP, Bachmann CM, et al (2020) Regulatory feedback cycle of the insulin-degrading enzyme and the amyloid precursor protein intracellular domain: implications for Alzheimer's disease. Aging Cell 19(11):e13264. https://pubmed.ncbi.nlm.nih.gov/33128835/

Lee SH, Kang J, Ho A, Watanabe H, Bolshakov VY, Shen J (2020) APP family regulates neuronal excitability and synaptic plasticity but not neuronal survival. Neuron 108(4):676–690.e8

Lefaucheur R, Fetter D, Gaillon G, Borden A, Maltête D (2014) Tendon rupture as an adverse effect of rasagiline. Clin Neuropharmacol 37(2):62

Lichtenthaler SF (2011) Alpha-secretase in Alzheimer's disease: molecular identity, regulation and therapeutic potential. J Neurochem 116:10–21

Little JT, Walsh S, Aisen PS (2008) An update on huperzine A as a treatment for Alzheimer's disease. Expert Opin Investig Drugs 17(2):209–215. https://pubmed.ncbi.nlm.nih.gov/18230054/

Lynch SY, Kaplow J, Zhao J, Dhadda S, Luthman J, Albala B (2018) P4-389: ELENBECESTAT, E2609, A BACE INHIBITOR: RESULTS FROM A PHASE-2 STUDY IN SUBJECTS WITH MILD COGNITIVE IMPAIRMENT AND MILD-TO-MODERATE DEMENTIA DUE TO ALZHEIMER'S DISEASE. Alzheimers Dement 14(7S_Part_31):P1623

Maher-Edwards G, De'Ath J, Barnett C, Lavrov A, Lockhart A (2015) A 24-week study to evaluate the effect of rilapladib on cognition and cerebrospinal fluid biomarkers of Alzheimer's disease. Alzheimers Dement 1(2):131–140

Mancini F, De Simone A, Andrisano V (2011) Beta-secretase as a target for Alzheimer's disease drug discovery: an overview of in vitro methods for characterization of inhibitors. Anal Bioanal Chem 400:1979–1996

Mathew B, Oh JM, Baty RS, Batiha GE, Parambi DG, Gambacorta N, Nicolotti O, Kim H (2021) Piperazine-substituted chalcones: a new class of MAO-B, AChE, and BACE-1 inhibitors for the treatment of neurological disorders. Environ Sci Pollut Res 28:38855–38866

Matthews DC, Ritter A, Thomas RG, Andrews RD, Lukic AS, Revta C et al (2021) Rasagiline effects on glucose metabolism, cognition, and tau in Alzheimer's dementia. Alzheimers Dement 7(1):e12106

Menard C, Bastianetto S, Quirion R (2013) Neuroprotective effects of resveratrol and epigallocatechin gallate polyphenols are mediated by the activation of protein kinase C gamma. Front Cell Neurosci 7:281

Miners JS, Baig S, Palmer J, Palmer LE, Kehoe PG, Love S (2008) Abeta-degrading enzymes in Alzheimer's disease. Brain Pathol 18(2):240–252. https://pubmed.ncbi.nlm.nih.gov/18363935/

Mohamed T, Zhao X, Habib LK, Yang J, Rao PPN (2011) Design, synthesis and structure-activity relationship (SAR) studies of 2,4-disubstituted pyrimidine derivatives: dual activity as cholinesterase and Aβ-aggregation inhibitors. Bioorg Med Chem 19(7):2269–2281

Moussa-Pacha NM, Abdin SM, Omar HA, Alniss H, Al-Tel TH (2020) BACE1 inhibitors: current status and future directions in treating Alzheimer's disease. Med Res Rev 40(1):339–384. https://pubmed.ncbi.nlm.nih.gov/31347728/

Musial A, Bajda M, Malawska B (2007) Recent developments in cholinesterases inhibitors for Alzheimer's disease treatment. Curr Med Chem 14(25):2654–2679

Nalivaeva NN, Belyaev ND, Lewis DI, Pickles AR, Makova NZ, Bagrova DI et al (2012) Effect of sodium valproate administration on brain neprilysin expression and memory in rats. J Mol Neurosci 46(3):569–577

Nalivaeva NN, Zhuravin IA, Turner AJ (2020) Neprilysin expression and functions in development, ageing and disease. Mech Ageing Dev 192:111363. https://pubmed.ncbi.nlm.nih.gov/32987038/

O'Brien RJ, Wong PC (2011) Amyloid precursor protein processing and Alzheimer's disease. Annu Rev Neurosci 34:185–204

Oddo S, Caccamo A, Kitazawa M, Tseng BP, LaFerla FM (2003) Amyloid deposition precedes tangle formation in a triple transgenic model of Alzheimer's disease. Neurobiol Aging 24(8):1063–1070

Oefner C, D'Arcy A, Hennig M, Winkler FK, Dale GE (2000) Structure of human neutral endopeptidase (neprilysin) complexed with phosphoramidon. J Mol Biol 296(2):341–349. http://www.idealibrary.com

Oefner C, Pierau S, Schulz H, Dale GE (2007) Structural studies of a bifunctional inhibitor of neprilysin and DPP-IV. Acta Crystallogr D Biol Crystallogr 63(9):975–981

Pacheco-Quinto J, Herdt A, Eckman CB, Eckman EA (2013) Endothelin-converting enzymes and related metalloproteases in Alzheimer's disease. J Alzheimers Dis 33(Suppl 1):S101. https://pubmed.ncbi.nlm.nih.gov/22903130/

Palmer JC, Kehoe PG, Love S (2010) Endothelin-converting enzyme-1 in Alzheimer's disease and vascular dementia. Neuropathol Appl Neurobiol 36(6):487–497. https://pubmed.ncbi.nlm.nih.gov/20345647/

Payne A, Nahashon S, Taka E, Adinew GM, Soliman KFA (2022) Epigallocatechin-3-Gallate (EGCG): new therapeutic perspectives for neuroprotection, aging, and neuroinflammation for the modern age. Biomol Ther 12:371

Pérez-Areales FJ, Betari N, Viayna A, Pont C, Espargaró A, Bartolini M et al (2017) Design, synthesis and multitarget biological profiling of second-generation anti-Alzheimer rhein-huprine hybrids. Future Med Chem 9(10):965–981

Pivovarova O, Höhn A, Grune T, Pfeiffer AFH, Rudovich N (2016) Insulin-degrading enzyme: new therapeutic target for diabetes and Alzheimer's disease? Ann Med 48:614–624

Placanica L, Zhu L, Yue-Ming L (2009) Gender- and age-dependent γ-secretase activity in mouse brain and its implication in sporadic Alzheimer disease. PLoS One 4(4):e5088

Postina R (2012) Activation of α-secretase cleavage. J Neurochem 120:46–54

Pradeep S, Jain AS, Dharmashekara C, Prasad SK, Kollur SP, Syed A, Shivamallu C (2020) Alzheimer's disease and herbal combination therapy: a comprehensive review. J Alzheimers Dis Rep 4(1):417–429
Prati F, De Simone A, Bisignano P, Armirotti A, Summa M, Pizzirani D et al (2015) Multitarget drug discovery for Alzheimer's disease: triazinones as BACE-1 and GSK-3β inhibitors. Angew Chem Int Ed 54(5):1578–1582
Prati F, Bottegoni G, Bolognesi ML, Cavalli A (2018) BACE-1 inhibitors: from recent single-target molecules to multitarget compounds for Alzheimer's disease. J Med Chem 61:619–637
Prinzen C, Müller U, Endres K, Fahrenholz F, Postina R (2005) Genomic structure and functional characterization of the human ADAM10 promoter. FASEB J 19(11):1522–1524
Qiu WQ, Folstein MF (2006) Insulin, insulin-degrading enzyme and amyloid-beta peptide in Alzheimer's disease: review and hypothesis. Neurobiol Aging 27(2):190–198. https://pubmed.ncbi.nlm.nih.gov/16399206/
Rao AA, Sridhar GR, Das UN (2007) Elevated butyrylcholinesterase and acetylcholinesterase may predict the development of type 2 diabetes mellitus and Alzheimer's disease. Med Hypotheses 69(6):1272–1276
Rosini M, Simoni E, Caporaso R, Minarini A (2016) Multitarget strategies in Alzheimer's disease: benefits and challenges on the road to therapeutics. Future Med Chem 8:697–711
Sasidharakurup H, Pradeep M, Bhaskaran M, Priya A, Pradeep E, Kadavath S, Nair B, Diwakar S (2018) Modeling of glutamate pathway in Alzheimer's disease using biochemical systems theory. In: 2018 international conference on advances in computing, communications and informatics (ICACCI). IEEE, Piscataway, pp 650–654
Sathya M, Premkumar P, Karthick C, Moorthi P, Jayachandran KS, Anusuyadevi M (2012) BACE1 in Alzheimer's disease. Clin Chim Acta 414:171–178. https://pubmed.ncbi.nlm.nih.gov/22926063/
Schneider LS, Geffen Y, Rabinowitz J, Thomas RG, Schmidt R, Ropele S et al (2019) Low-dose ladostigil for mild cognitive impairment: a phase 2 placebo-controlled clinical trial. Neurology 93(15):e1474–e1484
Seckl JR (2010) 11b-hydroxysteroid dehydrogenases. Stress science: neuroendocrinology, p 179
Sharma S, Behl T, Kumar A, Sehgal A, Singh S, Sharma N et al (2021) Targeting endothelin in Alzheimer's disease: a promising therapeutic approach. Biomed Res Int 2021:7396580. https://pubmed.ncbi.nlm.nih.gov/34532504/
Silva T, Reis J, Teixeira J, Borges F (2014) Alzheimer's disease, enzyme targets and drug discovery struggles: from natural products to drug prototypes. Ageing Res Rev 15(1):116–145. https://pubmed.ncbi.nlm.nih.gov/24726823/
Soria Lopez JA, González HM, Léger GC (2019) Alzheimer's disease. In: Handbook of clinical neurology. Elsevier B.V., Amsterdam, pp 231–255
Sousa L, Guarda M, Meneses MJ, Macedo MP, Vicente MH (2021) Insulin-degrading enzyme: an ally against metabolic and neurodegenerative diseases. J Pathol 255(4):346–361. https://pubmed.ncbi.nlm.nih.gov/34396529/
Steiner H, Fukumori A, Tagami S, Okochi M (2018) Making the final cut: pathogenic amyloid-β peptide generation by γ-secretase. Cell Stress 2(11):292. /pmc/articles/PMC6551803/
Sun MK, Alkon DL (2006) Bryostatin-1: pharmacology and therapeutic potential as a CNS drug. CNS Drug Rev 12(1):1–8
Talesa VN (2001) Acetylcholinesterase in Alzheimer's disease. Mech Ageing Dev 122(16):1961–1969
Tamagno E, Guglielmotto M, Aragno M, Borghi R, Autelli R, Giliberto L et al (2008) Oxidative stress activates a positive feedback between the γ- and β-secretase cleavages of the β-amyloid precursor protein. J Neurochem 104(3):683–695
Taylor HA, Przemylska L, Clavane EM, Meakin PJ (2022) BACE1: more than just a β-secretase. Obes Rev 23(7):e13430. https://pubmed.ncbi.nlm.nih.gov/35119166/
Tesco G, Koh YH, Kang EL, Cameron AN, Das S, Sena-Esteves M et al (2007) Depletion of GGA3 stabilizes BACE and enhances β-secretase activity. Neuron 54(5):721–737

Tolia A, De Strooper B (2009) Structure and function of gamma-secretase. Semin Cell Dev Biol 20(2):211–218. https://pubmed.ncbi.nlm.nih.gov/19007897/

Vassar R, Kovacs DM, Yan R, Wong PC (2009) The β-secretase enzyme BACE in health and Alzheimer's disease: regulation, cell biology, function, and therapeutic potential. J Neurosci 29:12787–12794

Vormfelde SV, Pezous N, Lefèvre G, Kolly C, Neumann U, Jordaan P et al (2020) A pooled analysis of three randomized phase I/IIa clinical trials confirms absence of a clinically relevant effect on the QTc interval by Umibecestat. Clin Transl Sci 13(6):1316–1326

Wang XX, Tan MS, Yu JT, Tan L (2014) Matrix metalloproteinases and their multiple roles in Alzheimer's disease. Biomed Res Int 2014:908636

Wang ZB, Wang ZT, Sun Y, Tan L, Yu JT (2022) The future of stem cell therapies of Alzheimer's disease. Ageing Res Rev 80:101655

Webster SP, McBride A, Binnie M, Sooy K, Seckl JR, Andrew R et al (2017) Selection and early clinical evaluation of the brain-penetrant 11β-hydroxysteroid dehydrogenase type 1 (11β-HSD1) inhibitor UE2343 (XanamemTM). Br J Pharmacol 174(5):396–408. https://pubmed.ncbi.nlm.nih.gov/28012176/

Weinreb O, Amit T, Bar-Am O, Youdim MBH, Center ET (2016) Themed section: updating neuropathology and neuropharmacology of monoaminergic systems neuroprotective effects of multifaceted hybrid agents targeting MAO, cholinesterase, iron and β-amyloid in ageing and Alzheimer's disease. Br J Pharmacol 173:2080–2094

Wen Y, Yu WH, Maloney B, Bailey J, Ma J, Marié I et al (2008) Transcriptional regulation of β-secretase by p25/cdk5 leads to enhanced amyloidogenic processing. Neuron 57(5):680–690

Wozniak MA, Itzhaki RF, Shipley SJ, Dobson CB (2007) Herpes simplex virus infection causes cellular β-amyloid accumulation and secretase upregulation. Neurosci Lett 429(2–3):95–100

Xiong K, Cai H, Luo XG, Struble RG, Clough RW, Yan XX (2007) Mitochondrial respiratory inhibition and oxidative stress elevate β-secretase (BACE1) proteins and activity in vivo in the rat retina. Exp Brain Res 181(3):435–446

Yiannopoulou KG, Papageorgiou SG (2013) Current and future treatments for Alzheimer's disease. Ther Adv Neurol Disord 6(1):19–33. https://doi.org/10.1177/1756285612461679

Yu M, Liu Y, Shen J, Lv D, Zhang J (2016) Meta-analysis of BACE1 gene rs638405 polymorphism and the risk of Alzheimer's disease in Caucasion and Asian population. Neurosci Lett 616:189–196

Yu T, Wan F, Liu C, Zhang X, Liu Z, Zhang J et al (2020) Asparagine endopeptidase inhibitor protects against fenpropathrin-induced neurodegeneration via suppressing α-synuclein aggregation and neuroinflammation. Eur J Pharmacol 888:173586

Yu TW, Lane HY, Lin CH (2021) Novel therapeutic approaches for Alzheimer's disease: an updated review. Int J Mol Sci 22(15):8208. https://pubmed.ncbi.nlm.nih.gov/34360973/

Zha X, Lamba D, Zhang L, Lou Y, Xu C, Kang D et al (2016) Novel tacrine-benzofuran hybrids as potent multitarget-directed ligands for the treatment of Alzheimers disease: design, synthesis, biological evaluation, and X-ray crystallography. J Med Chem 59(1):114–131

Zhang Z, Xie M, Ye K (2016) Asparagine endopeptidase is an innovative therapeutic target for neurodegenerative diseases. Expert Opin Ther Targets 20:1237–1245

Zhang P, Xu S, Zhu Z, Xu J (2019) Multi-target design strategies for the improved treatment of Alzheimer's disease. Eur J Med Chem 176:228–247

Zhou S, Zhou H, Walian PJ, Jap BK (2007) Regulation of γ-secretase activity in Alzheimer's disease. Biochemistry 46:2553–2563

Zhu Y, Xiao K, Ma L, Xiong B, Fu Y, Yu H et al (2009) Design, synthesis and biological evaluation of novel dual inhibitors of acetylcholinesterase and β-secretase. Bioorg Med Chem 17(4):1600–1613

Role of Enzymes in Amyotrophic Lateral Sclerosis

3

S. Maya and M. Shamnas

Abstract

Amyotrophic lateral sclerosis (ALS) refers to a deadly neurodegenerative disease where neurons are selectively damaged or destroyed in both the spinal cord and brain. The precise cause of ALS is uncertain, but scientists have been able to establish that it involves some biochemical disturbances and enzyme disorders. The involvement of specific enzymes in the etiology of ALS requires extensive elucidation. It is important to provide further details about the complicated interconnection among enzyme mismanagement and the deterioration process in ALS, as it may help in developing a cure or intervention for the debilitating consequence of the disease.

Keywords

Amyotrophic lateral sclerosis · Neurodegeneration · Superoxide dismutase · Glutamate transporters · Proteolytic enzymes · Metabolic enzymes

3.1 Introduction

Amyotrophic lateral sclerosis (ALS), or Lou Gehrig's disease, is considered an idiopathic, fatal neurological disease that affects motor neurons. The progressive degeneration of motor neurons in the brain and spinal cord affects the control of voluntary muscle movements. As a result, the movements of voluntary muscles become impaired, leading to muscle weakness and atrophy. Eventually, the person with ALS

S. Maya
Department of Pharmacology, KMCT College of Pharmacy, Malappuram, Kerala, India

M. Shamnas (✉)
Department of Pharmacy Practice, KMCT College of Pharmacy, Malappuram, Kerala, India

B. Mathew, D. G. T. Parambi (eds.), *Enzymes in Neurodegenerative Disorders*,
https://doi.org/10.1007/978-981-97-6822-6_3

dies from respiratory failure (Bennett et al. 2019; Brown 2018). The research on the specific case suggests that multiple factors are responsible for the development and progression of the disease, including genetic, environmental toxins, clinical factors, demographic factors, and the patient's nutritional status (Andrew and Charles 1993; Barber et al. 2006). However, the interplay between multiple mechanisms causes damage to critical cellular processes and initiates degeneration pathways. The disease may take several forms, such as sporadic ALS with no family history and familial ALS with a clear genetic component (Siddique and Siddique 2008).

Several studies suggest that various enzyme activities and expressions are altered in ALS and may be key determinants for its pathogenesis. Enzymes like these control different functions including oxidative stress, protein breakdown, and nucleic acid generation. Also, malfunctioning enzymes that play a role in mitochondrial energy generation and glutamate metabolism are evident in ALS (Gandhi and Abramov 2012; Rothstein et al. 1990). In this chapter, we discuss the pivotal role of enzymes in ALS and explore their impact on disease development and progression. Identification of enzyme dysfunction in ALS has led to the search for possible drug targets and biomarkers in the disease.

3.2 Enzymes and Their Significance in ALS

In the context of ALS, certain enzymes have been associated with key factors contributing to the development and progression of the condition.

3.2.1 Superoxide Dismutase 1 (SOD1)

SOD1 is an enzyme essential for the body's antioxidant defence system, neutralising harmful reactive oxygen species (ROS) produced during cellular metabolism. It plays a crucial role in ALS, with mutations in the SOD1 gene affecting approximately 20% of familial cases (Barbeito et al. 2004; Brown and Robberecht 2001). These mutations cause faulty SOD1 proteins to misfold and aggregate within motor neurons, which control voluntary muscle movements (Kim et al. 2016). This accumulation can lead to toxic protein aggregates, which interfere with normal function and cause degeneration and death of motor neurons. The exact mechanism behind mutant SOD1's degeneration is not fully understood, but it involves oxidative stress, mitochondrial dysfunction, excitotoxicity, and impairment of cellular pathways involved in protein degradation and quality control (Barber et al. 2006; Barber and Shaw 2010; Bowling et al. 1993).

The interplay between these mechanisms contributes to the progressive degeneration of motor neurons in ALS. As mutant SOD1 aggregates interfere with cellular processes, they disrupt protein quality control mechanisms and intracellular transport pathways, potentially contributing to oxidative stress within motor neurons (Baillet et al. 2010; Bergeron 1995; Chen et al. 2021). Inadequate antioxidant defences result in the generation of reactive oxygen species (ROS).

The ROS formation is described as follows:

1. Dismutation of superoxide to oxygen and hydrogen peroxide (Rakhit and Chakrabartty 2006; Wright et al. 2019).

$$2O_2^- + 2H^+ \rightarrow O_2 + H_2O_2$$

2. SOD1-catalyzed dismutation of superoxide reduces and oxidises the copper centre (Wright et al. 2019).
   (a) $O_2^- Cu^{2+}SOD1 \rightarrow O_2 + Cu^{1+}SOD1$
   (b) $O_2^- + 2H^+ + Cu^{1+}SOD1 \rightarrow H_2O_2 + Cu^{2+}SOD1$

These impaired cellular events can lead to mitochondrial dysfunction (Cunha-Oliveira et al. 2020). Mitochondrial dysfunction can lead to an imbalance in calcium homeostasis, which can release calcium into the cytoplasm, disrupting normal calcium signalling and regulation (Rossi et al. 2013; Kawamata and Manfredi 2010). This increase in intracellular calcium triggers the fusion of glutamate-containing synaptic vesicles within the presynaptic membrane, releasing glutamate into the synaptic cleft, the small space between the presynaptic neuron and the postsynaptic neuron. The dysregulation of calcium homeostasis in ALS disrupts glutamate's regular release and support at synapses. Excessive calcium influx can lead to an excessive release of glutamate at the synapses between neurons, exacerbating excitotoxicity (Bosch et al. 2006). This prolonged and excessive glutamate release and excitotoxicity can lead to impaired cellular functions, cellular damage and eventually motor neuron degeneration (Peggion et al. 2022). The dysfunctional mitochondria produce more ROS, leading to further calcium dysregulation and excitotoxicity, creating a harmful feedback loop.

The SOD1 dimer structure is the most studied ALS mutant, consisting of two identical subunits (Saccon et al. 2013; Thomas et al. 1974). Over 200 mutations have been reported in the SOD1 gene and are associated with familial ALS cases (Gagliardi et al. 2023). Each modification can have distinct effects on the protein's structure and function, leading to the toxic loss of function observed in ALS (Hough et al. 2004). Studying the locations and consequences of these mutations within the SOD1 dimer structure is crucial for understanding the molecular mechanisms of ALS pathogenesis and developing targeted therapies to address the disease (Wright et al. 2019).

3.2.2 Glutamate Transporters

Glutamate transporters are membrane-bound proteins regulating neurotransmitter levels in the central nervous system, including the brain and spinal cord. They maintain proper neurotransmission and prevent excitotoxicity. Deficiencies in these transporters lead to glutamate accumulation, causing overstimulation of neurons and neurotoxicity (Malik and Willnow 2019; Pajarillo et al. 2019). EAAT1 (GLAST) and EAAT2 (GLT) are two main subtypes relevant to ALS (Andrew and Charles 1993).

3.2.2.1 Excitatory Amino Acid Transporter 1 (EAAT1) or Glutamate-Aspartate Transporter (GLAST)

EAAT1 is primarily expressed in astrocytes, non-neuronal cells that support and protect neurons in the central nervous system (Barbeito et al. 2004). It is particularly abundant in regions like the cerebellum and brainstem. It plays a crucial role in regulating extracellular levels of the neurotransmitter glutamate (Parkin et al. 2018). After the glutamate release by neurons into the synaptic cleft, EAAT1 actively transports glutamate back into astrocytes, effectively clearing it from the synaptic space. This process helps terminate synaptic signalling and prevent excessive activation of glutamate receptors, thus avoiding excitotoxicity (Malik and Willnow 2019). This neuroprotective role is critical for maintaining neuronal health and preventing ALS.

In ALS, dysfunction of EAAT1 in astrocytes can lead to impaired glutamate clearance, contributing to an accumulation of extracellular glutamate and excitotoxicity, particularly affecting motor neurons. Eventually, these events will lead to motor neuron degeneration and contribute to the progression of ALS (Malik and Willnow 2019; Parkin et al. 2018).

3.2.2.2 Excitatory Amino Acid Transporter 2 (EAAT2, Also Known as GLT-1)

EAAT2 is also mainly expressed in astrocytes and is the most abundant glutamate transporter in the central nervous system. Like EAAT1, EAAT2 is responsible for clearing glutamate from the synaptic cleft (Fig. 3.1), preventing excessive activation of glutamate receptors (Malik and Willnow 2019).

In ALS, the expression and function of EAAT2 can be reduced, leading to decreased glutamate clearance and increased excitotoxicity (Rossi et al. 2013). Excessive glutamate overstimulates glutamate receptors, leading to calcium influx and subsequent cellular damage. Excitotoxicity triggers the production of ROS and oxidative stress within motor neurons. Elevated ROS levels can cause cellular damage and contribute to neurodegeneration (Barnham et al. 2004; Parkin et al. 2018).

3.2.3 Proteolytic Enzymes

Proteolytic enzymes, also known as proteases, play a role in various cellular processes, including protein degradation, regulation of cell signalling, and maintaining cellular homeostasis (Antalis and Buzza 2016).

In the context of ALS, proteolytic enzymes are involved in several mechanisms that contribute to the pathogenesis of neurodegeneration.

3.2.3.1 Protein Aggregation and Misfolding

Proteolytic enzymes break down misfolded or damaged proteins and help maintain cellular protein quality control. However, in ALS, the dysfunction of proteolytic systems can contribute to accumulating misfolded proteins and forming toxic protein aggregates (Abramov et al. 2020; Szelechowski et al. 2018). In ALS, the

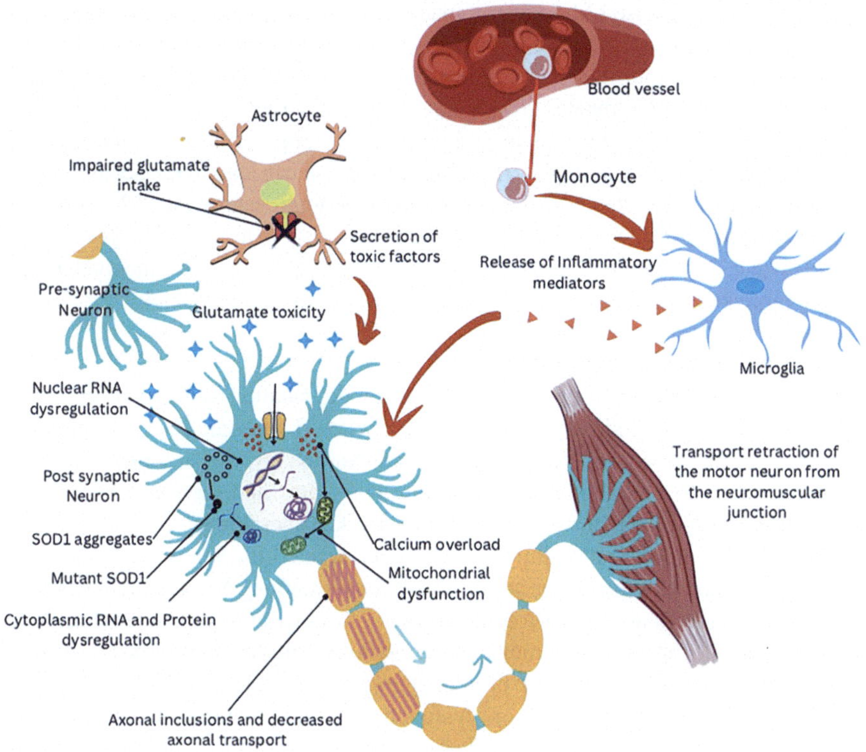

Fig. 3.1 Possible mechanisms for SOD1-related ALS

aggregation-prone mutant SOD1 proteins and TDP-43 (TAR DNA-binding protein 43, crucial for several cellular processes, including gene expression regulation, RNA metabolism, and protein homeostasis) can overwhelm the cellular protein quality control systems (Iwashima et al. 2010; Kametani et al. 2016).

(a) **Chaperones**: Molecular chaperones, also known as heat shock proteins (HSPs), play a critical role in maintaining protein homeostasis and ensuring proper protein folding within cells (Chen et al. 2011). They assist in folding, preventing misfolding, and refolding denatured or misfolded proteins (Brown 2007; Camberg et al. 2013). Chaperones transport proteins within the cell, guiding them to their correct subcellular locations. This is central to proteostasis, the cellular balance between protein synthesis, folding, trafficking, and degradation, ensuring protein solubility and functionality (Chen et al. 2011).
(b) **Protein Disulfide Isomerases (PDIs)**: PDIs are a crucial enzyme in protein folding and quality control, primarily forming and rearranging disulfide bonds between cysteine residues. PDIs are located in the extracellular matrix (ER) and help proteins fold correctly in challenging environments. ER stress, like

Alzheimer's disease (ALS), can impact PDI function and disrupt protein folding, potentially contributing to misfolding and aggregation (Walker 2010).

(c) **Calpains**: Calcium-dependent proteases that influence the formation and clearance of protein aggregate. Dysregulated calpain activity can compromise cellular proteostasis, the balance between protein synthesis, folding, and degradation (Stifanese et al. 2014). This disruption can lead to the accumulation of misfolded proteins, which are more likely to aggregate in ALS (Ruegsegger and Saxena 2016). Some protein aggregates may serve as substrates for calpain cleavage. This can lead to the generation of smaller protein fragments that can be more susceptible to aggregation or contribute to cellular toxicity.

3.2.3.2 Autophagy Impairment

Autophagy is a cellular process that involves the degradation and recycling of cellular components, including proteins and organelles (Lee et al. 2015). Proteolytic enzymes within lysosomes play a crucial role in autophagy, and autophagy impairment results in the accumulation of protein aggregates and damaged organelles, contributing to disease pathogenesis (Levine and Kroemer 2008).

Autophagosomes engulf cellular components targeted for degradation, and proteolytic enzymes play a role in autophagosome formation. Autophagosomes fuse with lysosomes containing proteolytic enzymes, forming autolysosomes (Fig. 3.2). This clearance of proteins is crucial in maintaining cellular homeostasis and preventing the accumulation of toxic protein aggregates (Levine and Kroemer 2008).

Mitophagy targets damaged or dysfunctional mitochondria for degradation (Martinez-Vicente 2017). Proteolytic enzymes in lysosomes break down the engulfed mitochondria, preventing the release of harmful molecules and supporting cellular health. Dysfunction of lysosomal proteolytic enzymes can result in undegraded materials within autolysosomes, disrupting autophagy and leading to the buildup of damaged proteins and organelles. Impaired autophagy may result in a reduced supply of substrates for lysosomal degradation, affecting overall cellular clearance mechanisms (Levine and Kroemer 2008).

Proteolytic enzymes, specifically cathepsins, have been implicated in the degradation of misfolded proteins and the clearance of cellular debris during autophagy (Wootz et al. 2006). Dysfunction in these enzymes can hinder proper protein turnover and disposal, contributing to the formation of protein aggregates observed in ALS patients (Chen et al. 2012; Chua et al. 2022).

3.2.3.3 Ubiquitin-Proteasome System (UPS) Dysfunction

The UPS maintains protein homeostasis by breaking down targeted proteins marked with ubiquitin tags (Quinet et al. 2020). Ubiquitination involves attaching ubiquitin to lysine residues on target proteins, signalling the protein's recognition by the proteasome for degradation (Haouari et al. 2022). The multi-subunit complex proteasome recognizes ubiquitinated proteins, unfolds them, and cleaves them into smaller peptides. The proteolytic activity is essential for destroying short-lived and damaged proteins.

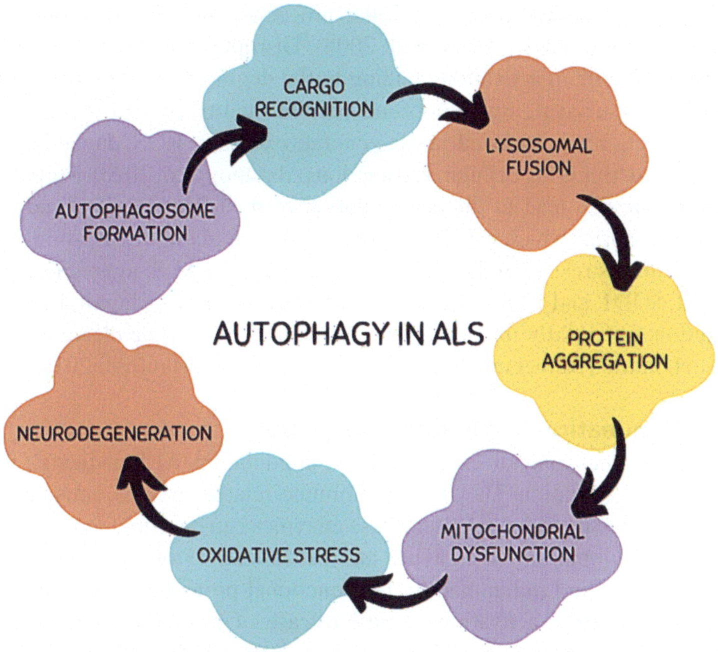

Fig. 3.2 Flow chart for autophagy in ALS

In neurodegenerative diseases like ALS, changes in ubiquitin-proteasome signalling can lead to dysregulated ubiquitination, resulting in aberrant ubiquitin chain structures or alterations in ubiquitin-conjugating enzyme levels (Bendotti et al. 2012; Petrucelli and Dawson 2004). Impaired proteasomal components or decreased activity can result from genetic mutations, oxidative stress, or other factors. Deficient proteolytic enzymes within the proteasome cleave ubiquitinated substrates, leading to protein accumulation and aggregates, which are toxic to cells and contribute to neurodegenerative diseases (Bett 2016; Huang and Figueiredo-Pereira 2010).

3.2.3.4 Mitochondrial Dysfunction

Mitochondria are crucial organelles responsible for energy production, calcium regulation, and cell survival. Dysfunctional mitochondria can lead to oxidative stress, energy deficits, and neuronal damage in ALS. Proteolytic enzymes are crucial in maintaining mitochondrial quality control by degrading damaged mitochondria (Guo et al. 2013).

In ALS, impaired proteolysis can result in the accumulation of dysfunctional mitochondria, producing reactive oxygen species and contributing to oxidative stress (Gandhi and Abramov 2012). Mitophagy is triggered by the ubiquitination of damaged mitochondrial proteins, which are then degraded by proteasomes and lysosomes. Impaired lysosomal function can disrupt autophagic processes, resulting in the accumulation of damaged mitochondria (Fröhlich and Montoro 2023).

Calpains play essential roles in cellular processes, and abnormal activation of calpains can result in ALS (Vosler et al. 2008). Disrupted calcium homeostasis can cause abnormal calpain activation, leading to the degradation of essential proteins involved in cell survival, signalling, and protein quality control (Stifanese et al. 2014). Dysfunctional mitochondria can generate excessive ROS, damaging cellular components and triggering cellular dysfunction (Abramov et al. 2020). Mitochondrial dysfunction can also lead to the accumulation of misfolded or damaged proteins within mitochondria, which can escape expected proteolytic degradation and contribute to cellular stress (Gandhi and Abramov 2012). In ALS, aggregated proteins like mutant SOD1 and TDP-43 can be associated with mitochondria and impair their function, potentially impacting mitophagy and proteolytic clearance of damaged mitochondria (Kametani et al. 2016; Kawamata and Manfredi 2010).

3.2.3.5 Inflammation and Immune Response

Proteolytic enzymes play a crucial role in modulating ALS patients' immune response and inflammation. They process immune-related proteins, activate immune cells, and influence cytokine release. These enzymes can contribute to neuroinflammation and inflammation by cleaving and processing cytokines, which regulate immune responses and inflammation. Dysfunctional proteolytic enzymes can lead to abnormal microglial activation and the release of pro-inflammatory factors, a hallmark of ALS (Almer et al. 2001; Lu et al. 2016). Inflammation and immune responses can produce ROS, leading to oxidative stress. Proteolytic enzymes activate immune cells and release ROS and cytokines, which can lead to a chronic inflammatory environment in the central nervous system (Béland et al. 2020).

Plasmin is a serine protease that plays a crucial role in fibrinolysis, which involves breaking down fibrin clots to maintain blood vessel patency. Beyond its role in fibrinolysis, plasmin is also involved in extracellular matrix remodelling, wound healing, and tissue repair (Lin et al. 2020). Dysregulated plasmin activity can lead to excessive degradation of extracellular matrix components, releasing bioactive molecules that contribute to inflammation. This inflammation can exacerbate neurodegeneration and tissue damage in ALS (Yepes et al. 2021).

3.2.3.6 Axonal Degeneration

Proteolytic enzymes play a crucial role in axonal degeneration, an essential aspect of ALS pathology. They degrade axonal cytoskeletal proteins and components, disrupting axonal transport and stability. Dysfunctional proteolysis can disturb the balance of axonal transport, leading to mislocalised proteins and axonal degeneration during stress or injury (Coleman 2022). Proteolytic enzymes break down axonal components and dismantle the degenerating structures. Proper proteostasis within axons is essential for preventing the accumulation of misfolded or aggregated proteins, ensuring proper function and survival (Fischer and Glass 2007).

3.2.3.7 Blood-Brain Barrier (BBB) Integrity

Proteases can disrupt the BBB, which separates the central nervous system from the bloodstream. Increased plasmin activity can degrade BBB components, compromising their integrity (Niego and Medcalf 2014). Disruption could allow immune cells and harmful molecules into the central nervous system, contributing to neuroinflammation (Garbuzova-Davis et al. 2007, 2012). Tight junctions between endothelial cells form the BBB, and proteolytic enzymes can modulate their structure and stability. The basement membrane, a specialised extracellular matrix, is degraded by proteolytic enzymes, leading to its remodelling and potential BBB disruption (Rempe et al. 2016). Inflammation in ALS can trigger the release and activation of proteases, leading to the degradation of extracellular matrix proteins and basement membrane components (Maguire 2017).

Proteases like matrix metalloproteinases can directly degrade BBB proteins and extracellular matrix molecules, disrupting BBB (Łukaszewicz-Zając et al. 2014). Oxidative stress in ALS can also activate proteolytic enzymes, increasing permeability and potential immune cell infiltration. Activated astrocytes also release proteolytic enzymes that trigger BBB disruption during neuroinflammation (Evans et al. 2013).

The complex interplay between proteolytic enzymes, protein aggregation, autophagy, inflammation, and other cellular processes contributes to the development and progression of ALS (Rossi et al. 2013). Restoring the activity of proteolytic enzymes holds promise as a new therapeutic avenue for preventing or slowing down disease progression.

3.2.4 Metabolic Enzymes

Metabolic enzymes are essential for cellular energy production, biosynthesis, and metabolic pathway regulation. There are several clinical evidence suggesting the association of metabolic alterations with ALS (Nelson and Trotti 2022). In ALS patients, enzymes like creatine kinase and cytochrome oxidase show altered activity, leading to metabolic dysfunction in glucose, protein homeostasis, lipid metabolism, and oxidative stress regulation (Blasco et al. 2020; Browne et al. 2002). Here, we are discussing some metabolic enzymes and their roles in ALS.

3.2.4.1 Glucose Metabolism

Metabolic enzymes critically regulate glucose metabolism, the primary energy source for cells. In ALS, impaired glucose metabolism leads to a compromised energy supply to essential motor neurons, making them susceptible to degeneration (Allen et al. 2019; Lee et al. 2022). Metabolic enzymes involved in glucose metabolism are as follows:

(a) **Hexokinase**: Hexokinase catalyses the phosphorylation of glucose to glucose-6-phosphate, the first step in glycolysis. Altered hexokinase activity can impact the rate at which cells utilise glucose for energy production (Roberts and Miyamoto 2015).

(b) **Phosphofructokinase (PFK)**: PFK catalyses glycolysis regulatory enzyme that converts fructose-6-phosphate to fructose-1,6-bisphosphate. Dysregulated PFK activity can affect the rate of glucose utilisation in cells (Yi et al. 2019).
(c) **Pyruvate kinase**: Pyruvate kinase converts phosphoenolpyruvate (PEP) to pyruvate during glycolysis. Changes in pyruvate kinase activity can influence the production of pyruvate, a critical metabolite in energy production (Israelsen and Heiden 2015).
(d) **Lactate Dehydrogenase (LDH)**: LDH catalyses the interconversion of pyruvate and lactate, an essential step in maintaining cellular redox balance. Dysregulation of LDH activity can affect lactate production and cellular energetics (Han et al. 2023).
(e) **Glucose-6-Phosphate Dehydrogenase (G6PD)**: G6PD is involved in the pentose phosphate pathway, which generates NADPH and ribose-5-phosphate for biosynthesis and antioxidant defence. Altered G6PD activity can influence cellular redox status (Yang et al. 2019).
(f) **Enolase**: Enolase is an enzyme that catalyses the conversion of 2-phosphoglycerate to PEP in glycolysis. Changes in enolase activity can affect the glycolytic pathway's efficiency (Fukano and Kimura 2014).
(g) **Glucose Transporters (GLUTs)**: GLUTs are transport proteins that facilitate glucose uptake into cells. Dysregulated GLUT expression or activity can impact glucose uptake in motor neurons and other cells (Wang et al. 2020).
(h) **Pyruvate Dehydrogenase (PDH) Complex**: PDH converts pyruvate to acetyl-CoA, a critical step linking glycolysis to the citric acid cycle (Krebs cycle). Dysfunction of the PDH complex can affect the efficiency of glucose metabolism (Park et al. 2018).
(i) **Citrate Synthase**: Citrate synthase is an enzyme involved in the citric acid (Krebs cycle) within mitochondria. Altered citrate synthase activity can affect the metabolism of glucose-derived pyruvate (Iacobazzi and Infantino 2014).

Dysfunction in any of these metabolic enzymes can lead to disruptions in glucose metabolism, resulting in energy deficits, oxidative stress, and impaired cellular functions in motor neurons and other central nervous system cells (Tefera et al. 2021).

3.2.4.2 Protein Homeostasis

Metabolic enzymes are essential for protein homeostasis, cellular health processes, and quality control. ALS is caused by protein aggregates in motor neurons, causing toxicity and cell death. Dysregulated metabolic enzymes disrupt protein degradation and clearance mechanisms, resulting in toxic aggregate accumulation and contributing to ALS pathogenesis (Blokhuis et al. 2013). Here are some key metabolic enzymes involved in protein homeostasis and their roles in ALS:

(a) **UPS**: While not an enzyme, the UPS is a critical cellular pathway for protein degradation. Proteasomal enzymes, such as proteasome subunits, are responsible for degrading misfolded and unwanted proteins. Dysregulation of the UPS can lead to the accumulation of misfolded proteins in ALS (George and Tepe 2021).

(b) **HSPs**: Heat shock proteins, including HSP70 and HSP90, are molecular chaperones that assist in protein folding and prevent protein misfolding and aggregation. Dysregulated HSP activity can contribute to protein misfolding and collection in ALS (Chen et al. 2011).

(c) **PDIs**: PDIs are involved in forming and rearranging disulfide bonds in proteins, crucial for proper protein folding. Dysfunctional PDIs can lead to protein misfolding and aggregation in ALS (Walker 2010).

(d) **Molecular Chaperones**: Various molecular chaperones, such as HSPs and T-complex proteins, facilitate the folding of newly synthesised proteins and assist in refolding misfolded proteins. Dysfunction of molecular chaperones can lead to protein misfolding in ALS (Brown 2007; Chen et al. 2011).

(e) **Autophagy-Related Proteins**: Autophagy is a catabolic cellular process that degrades damaged or unwanted proteins and organelles. Metabolic enzymes, such as those involved in regulating autophagy (e.g. Mammalian target of rapamycin (mTOR)), can impact the clearance of misfolded proteins in ALS (Lee et al. 2015).

(f) **Lysosomal Enzymes**: Lysosomal enzymes, including cathepsins, are involved in the degradation of proteins during autophagy. Dysfunctional lysosomal enzymes can impair the clearance of protein aggregates in ALS (Monaco and Fraldi 2020).

(g) **Calpains**: The calcium-dependent proteases that can cleave proteins, potentially influencing protein quality control and degradation. Dysregulated calpain activity can lead to protein degradation and cellular damage in ALS (Stifanese et al. 2014).

(h) **Proteostasis Regulators**: Enzymes involved in post-translational modifications, such as kinases and phosphatases, can modulate the activity of proteins involved in proteostasis regulation. Dysregulation of these enzymes can impact protein homeostasis in ALS (Cheng et al. 2011; Hu et al. 2003).

(i) **Oxidoreductases**: Enzymes involved in redox reactions can influence protein folding and stability (Prabhulkar et al. 2012). Oxidative stress every day in ALS can impact the activity of these enzymes and contribute to protein misfolding (Hosaka et al. 2021).

(j) **Ubiquitin Ligases**: Ubiquitin ligases tag proteins for degradation by the UPS or autophagy (Lee et al. 2015). Dysregulation of ubiquitin ligases can impair the clearance of misfolded proteins in ALS (Schmidt et al. 2021).

Disruptions in protein homeostasis, including the accumulation of misfolded proteins and protein aggregates, are characteristic of ALS pathology (Lambert-Smith et al. 2022).

3.2.4.3 Lipid Metabolism

Lipids, including fatty acids, cholesterol, and phospholipids, play significant roles in maintaining cellular integrity and signalling (Sadler et al. 2022). Metabolic enzyme dysfunction in ALS affects lipid metabolism, leading to altered lipid profiles and compromised membrane integrity (Alessenko et al. 2020; Agrawal et al. 2022; Chaves-Filho et al. 2019). Here are some key metabolic enzymes involved in lipid metabolism and their potential roles in ALS:

(a) **Fatty Acid Synthase (FASN)**: FASN synthesises long-chain fatty acids, a critical component of lipid metabolism (Jensen-Urstad and Semenkovich 2012). Dysregulated FASN activity can contribute to lipid imbalance and cellular stress in ALS (Agrawal et al. 2022).
(b) **Acetyl-CoA Carboxylase (ACC)**: ACC is an enzyme involved in fatty acid synthesis by converting acetyl-CoA into malonyl-CoA, a precursor for fatty acid production (Chandel 2021). Dysregulation of ACC can impact lipid synthesis in motor neurons (Ronowska et al. 2018).
(c) **Carnitine Palmitoyltransferase (CPT)**: CPT enzymes facilitate the transport of long-chain fatty acids into mitochondria for beta-oxidation (Sebastián et al. 2009). Altered CPT activity can affect fatty acid metabolism and energy production in ALS (Trabjerg et al. 2021).
(d) **Lipoprotein Lipase (LPL)**: LPL is involved in the breakdown of triglycerides in lipoproteins, releasing fatty acids for energy utilisation. Dysregulated LPL activity can affect lipid homeostasis (Olivecrona 2016).
(e) **Adipose Triglyceride Lipase (ATGL)**: ATGL is an enzyme responsible for triglyceride breakdown, releasing fatty acids for energy production. Changes in ATGL activity can influence lipid metabolism in ALS (Huijsman et al. 2009).
(f) **Phospholipases**: Phospholipases, such as phospholipase A2 (PLA2), are involved in breaking membrane phospholipids. Dysregulated phospholipase activity can impact membrane integrity and lipid signalling (Alessenko et al. 2020).
(g) **Sphingolipid Metabolizing Enzymes**: Enzymes involved in sphingolipid metabolism, including ceramides and sphingomyelinases, can influence lipid balance and cellular signalling. Alterations in sphingolipid metabolism have been implicated in ALS (Ariga et al. 1998; Pralhada Rao et al. 2013).
(h) **Cholesterol Biosynthesis Enzymes**: Enzymes involved in cholesterol biosynthesis, such as HMG-CoA reductase, can influence lipid composition and cellular membrane properties in motor neurons (Qian et al. 2022).
(i) **Lipid Transport Proteins**: Proteins involved in lipid transport, such as apolipoproteins and lipoprotein receptors, can affect lipid metabolism and cholesterol homeostasis in ALS (Trostchansky 2019).
(j) **Lipid Signalling Enzymes**: Enzymes involved in lipid signalling pathways, including phosphoinositide kinases and phospholipase C, can influence cellular signalling cascades that may be altered in ALS (Balla 2013; Raghu et al. 2019).

Disruptions in lipid metabolism contribute to motor neuron vulnerability and disease progression (Alessenko et al. 2023).

3.2.4.4 Oxidative Stress Regulation

Metabolic enzymes are vital in regulating oxidative stress, a process involving the imbalance between ROS production and antioxidant defence systems (Gandhi and Abramov 2012; Shukla et al. 2011). In ALS, impaired metabolic enzyme activity disrupts ROS detoxification, resulting in heightened oxidative stress levels. Prolonged oxidative stress contributes to motor neuron damage and accelerates disease progression (Abramov et al. 2020; Carrì et al. 2003). Here are some metabolic enzymes involved in oxidative stress regulation and their potential roles in ALS:

(a) **Superoxide Dismutase (SOD)**: SOD enzymes, including Cu/Zn-SOD (SOD1), Mn-SOD (SOD2), and extracellular SOD (SOD3), are critical antioxidants that convert superoxide radicals into less harmful molecules (Gandhi and Abramov 2012). Mutations in SOD1 are linked to familial ALS, where dysregulated SOD1 activity can increase oxidative stress (Discussed under Sect. 3.2.1).

(b) **Catalase**: Catalase is an enzyme that breaks down hydrogen peroxide into water and oxygen, reducing oxidative stress. Changes in catalase activity can influence redox balance in ALS (Gandhi and Abramov 2012).

(c) **Glutathione Peroxidase (GPX)**: This enzyme, including GPx1 and GPx4, reduces lipid peroxides and protects cell membranes from oxidative damage. Altered GPx activity can impact antioxidant defence mechanisms in ALS (Angelova et al. 2021).

(d) **Glutathione Reductase**: Glutathione reductase helps regenerate reduced glutathione (GSH), a critical cellular antioxidant. Dysregulated glutathione reductase activity can affect GSH levels and redox status in motor neurons (Ballatori et al. 2009).

(e) **Glutathione *S*-Transferases (GSTs)**: GSTs detoxify electrophilic compounds and protect cells from oxidative damage. Changes in GST activity can influence cellular responses to oxidative stress in ALS (Espinosa-Diez et al. 2015).

(f) **NAD(P)H: Quinone Oxidoreductase 1 (NQO1)**: NQO1 is an enzyme that plays a role in maintaining antioxidants such as ubiquinone (coenzyme Q). Dysregulated NQO1 activity can impact the cellular redox state (Lee et al. 2021).

(g) **Cytochrome P450 Enzymes**: Some cytochrome P450 enzymes are involved in the metabolism of xenobiotics and endogenous compounds that can generate oxidative stress. Changes in their activity can contribute to oxidative damage (Veith and Moorthy 2018).

(h) **G6PD**: G6PD is involved in producing NADPH, a critical cofactor for cellular antioxidant systems. Altered G6PD activity can impact NADPH availability for redox reactions (Stanton 2012).

(i) **Isocitrate Dehydrogenase (IDH)**: Mutations in IDH genes have been associated with ALS. IDH enzymes play roles in generating NADPH and maintaining cellular redox balance (Tommasini-Ghelfi et al. 2019).

(j) **Selenoproteins**: Selenoproteins, including selenoprotein P and S, are involved in various antioxidant and redox regulatory processes. Changes in selenoprotein function can impact oxidative stress responses (Zhang et al. 2020).

(k) **Heme Oxygenase-1 (HO-1)**: HO-1 is an enzyme involved in heme catabolism and has antioxidant and cytoprotective functions. Its expression is upregulated in response to oxidative stress (Consoli et al. 2021).

Dysregulation of these metabolic enzymes can disrupt cellular antioxidant defence mechanisms, leading to increased oxidative stress in ALS.

3.3 Enzyme-Based Therapeutic Approaches

Given the intimate connection between enzymes and ALS pathogenesis, targeting specific enzymes has emerged as a promising therapeutic strategy. Researchers are exploring approaches to enhance the activity of beneficial enzymes while inhibiting or regulating the aberrant ones. For instance, small molecules downregulating SOD1 activity are being investigated for their potential to slow disease progression.

3.4 Conclusion

In conclusion, enzymes play a multifaceted role in the complex pathophysiology of ALS. These cellular catalysts are intimately associated with the disease development and progression, from genetic mutations involving enzymes such as SOD1 to wider, more comprehensive enzymatic processes that regulate oxidative stress, protein homeostasis, or metabolic balance.

Enzymes in ALS research offer a promising avenue for therapeutic development. Targeting specific enzymatic processes, either to restore normal function or to counteract disease-related dysfunction, holds the potential for slowing or even halting the progression of this devastating disease. By comprehensively understanding the impact of enzymes on pathways related to oxidative stress, protein aggregation, and neuroinflammation, researchers aim to pave the way for effective treatments that can alleviate symptoms and slow disease progression. Further research and clinical trials are required to translate these scientific insights into tangible advancements for ALS patients worldwide.

References

Abramov AY, Potapova EV, Dremin VV, Dunaev AV (2020) Interaction of oxidative stress and misfolded proteins in the mechanism of neurodegeneration. Life (Basel) 10(7):101. https://doi.org/10.3390/life10070101

Agrawal I, Lim YS, Ng S-Y, Ling S-C (2022) Deciphering lipid dysregulation in ALS: from mechanisms to translational medicine. Transl Neurodegener 11(1):48. https://doi.org/10.1186/s40035-022-00322-0

Alessenko AV, Gutner UA, Nebogatikov VO, Shupik MA, Ustyugov AA (2020) The role of lipids in the pathogenesis of lateral amyotrophic sclerosis. Zh Nevrol Psikhiatr Im S S Korsakova 120(10):108. https://doi.org/10.17116/jnevro2020120101108

Alessenko AV, Gutner UA, Shupik MA (2023) Involvement of lipids in the pathogenesis of amyotrophic lateral sclerosis. Life (Basel) 13(2):510. https://doi.org/10.3390/life13020510

Allen SP, Hall B, Woof R, Francis L, Gatto N, Shaw AC et al (2019) C9orf72 expansion within astrocytes reduces metabolic flexibility in amyotrophic lateral sclerosis. Brain 142(12):3771–3790. https://doi.org/10.1093/brain/awz302

Almer G, Guégan C, Teismann P, Naini A, Rosoklija G, Hays AP et al (2001) Increased expression of the pro-inflammatory enzyme cyclooxygenase-2 in amyotrophic lateral sclerosis. Ann Neurol 49(2):176–185. https://doi.org/10.1002/1531-8249(20010201)49:2<176::aid-ana37>3.0.co;2-x

Andrew E, Charles K (1993) Pathogenic mechanisms in sporadic amyotrophic lateral sclerosis. Can J Neurol Sci 20(4):286–296. https://doi.org/10.1017/s0317167100048198

Angelova PR, Esteras N, Abramov AY (2021) Mitochondria and lipid peroxidation in the mechanism of neurodegeneration: finding ways for prevention. Med Res Rev 41(2):770–784. https://doi.org/10.1002/med.21712

Antalis TM, Buzza MS (2016) Extracellular: plasma membrane proteases—serine proteases. Encyclopedia Cell Biol:650–660. https://doi.org/10.1016/B978-0-12-394447-4.10076-8

Ariga T, Jarvis WD, Yu RK (1998) Role of sphingolipid-mediated cell death in neurodegenerative diseases. J Lipid Res 39(1):1–16. https://doi.org/10.1016/s0022-2275(20)34198-5

Baillet A, Chanteperdrix V, Trocmé C, Casez P, Garrel C, Besson G (2010) The role of oxidative stress in amyotrophic lateral sclerosis and Parkinson's disease. Neurochem Res 35(10):1530–1537. https://doi.org/10.1007/s11064-010-0212-5

Balla T (2013) Phosphoinositides: tiny lipids with giant impact on cell regulation. Physiol Rev 93(3):1019–1137. https://doi.org/10.1152/physrev.00028.2012

Ballatori N, Krance SM, Notenboom S, Shi S, Tieu K, Hammond CL (2009) Glutathione dysregulation and the etiology and progression of human diseases. Biol Chem 390(3):191–214. https://doi.org/10.1515/bc.2009.033

Barbeito LH, Pehar M, Cassina P, Vargas MR, Peluffo H, Viera L et al (2004) A role for astrocytes in motor neuron loss in amyotrophic lateral sclerosis. Brain Res Rev 47(1–3):263–274. https://doi.org/10.1016/j.brainresrev.2004.05.003

Barber SC, Shaw PJ (2010) Oxidative stress in ALS: key role in motor neuron injury and therapeutic target. Free Radic Biol Med 48(5):629–641. https://doi.org/10.1016/j.freeradbiomed.2009.11.018

Barber SC, Mead RJ, Shaw PJ (2006) Oxidative stress in ALS: a mechanism of neurodegeneration and a therapeutic target. Biochim Biophys Acta 1762(11–12):1051–1067. https://doi.org/10.1016/j.bbadis.2006.03.008

Barnham KJ, Masters CL, Bush AI (2004) Neurodegenerative diseases and oxidative stress. Nat Rev Drug Discov 3(3):205–214. https://doi.org/10.1038/nrd1330

Béland L-C, Markovinovic A, Jakovac H, De Marchi F, Bilic E, Mazzini L et al (2020) Immunity in amyotrophic lateral sclerosis: blurred lines between excessive inflammation and inefficient immune responses. Brain Commun 2(2):fcaa124. https://doi.org/10.1093/braincomms/fcaa124

Bendotti C, Marino M, Cheroni C, Fontana E, Crippa V, Poletti A et al (2012) Dysfunction of constitutive and inducible ubiquitin-proteasome system in amyotrophic lateral sclerosis: implication for protein aggregation and immune response. Prog Neurobiol 97(2):101–126. https://doi.org/10.1016/j.pneurobio.2011.10.001

Bennett SA, Tanaz R, Cobos SN, Torrente MP (2019) Epigenetics in amyotrophic lateral sclerosis: a role for histone post-translational modifications in neurodegenerative disease. Transl Res 204:19–30. https://doi.org/10.1016/j.trsl.2018.10.002

Bergeron C (1995) Oxidative stress: its role in the pathogenesis of amyotrophic lateral sclerosis. J Neurol Sci 129:81–84. https://doi.org/10.1016/0022-510x(95)00071-9

Bett JS (2016) Proteostasis regulation by the ubiquitin system. Essays Biochem 60(2):143–151. https://doi.org/10.1042/ebc20160001

Blasco H, Lanznaster D, Veyrat-Durebex C, Hergesheimer R, Vourch P, Maillot F et al (2020) Understanding and managing metabolic dysfunction in amyotrophic lateral sclerosis. Expert Rev Neurother 20(9):907–919. https://doi.org/10.1080/14737175.2020.1788389

Blokhuis AM, Groen EJN, Koppers M, van den Berg LH, Pasterkamp RJ (2013) Protein aggregation in amyotrophic lateral sclerosis. Acta Neuropathol 125(6):777–794. https://doi.org/10.1007/s00401-013-1125-6

Bosch LVD, Damme PV, Bogaert E, Robberecht W (2006) The role of excitotoxicity in the pathogenesis of amyotrophic lateral sclerosis. Biochim Biophys Acta 1762(11–12):1068–1082. https://doi.org/10.1016/j.bbadis.2006.05.002

Bowling AC, Schulz JB, Brown RH Jr, Beal MF (1993) Superoxide dismutase activity, oxidative damage, and mitochondrial energy metabolism in familial and sporadic amyotrophic lateral sclerosis. J Neurochem 61(6):2322–2325. https://doi.org/10.1111/j.1471-4159.1993.tb07478.x

Brown IR (2007) Heat shock proteins and protection of the nervous system. Ann N Y Acad Sci 1113(1):147–158. https://doi.org/10.1196/annals.1391.032

Brown RH (2018) Amyotrophic lateral sclerosis and other motor neuron diseases. In: Jameson JL, Fauci AS, Kasper DL, Hauser SL, Longo DL, Loscalzo J (eds) Access medicine, 20th edn. McGraw-Hill Education, New York. https://accessmedicine.mhmedical.com/content.aspx?bookid=2129&sectionid=192532519

Brown RH, Robberecht W (2001) Amyotrophic lateral sclerosis: pathogenesis. Semin Neurol 21(02):131–140. https://doi.org/10.1055/s-2001-15260

Browne SE, Bowling AC, Baik MJ, Gurney M, Brown RH Jr, Beal MF (2002) Metabolic dysfunction in familial, but not sporadic, amyotrophic lateral sclerosis. J Neurochem 71(1):281–287. https://doi.org/10.1046/j.1471-4159.1998.71010281.x

Camberg JL, Doyle SM, Johnson DM, Wickner S (2013) Molecular chaperones. In: Brenner's encyclopedia of genetics, vol 2. Elsevier, Amsterdam, pp 456–460. https://doi.org/10.1016/B978-0-12-374984-0.00221-7

Carrì MT, Ferri A, Cozzolino M, Calabrese L, Rotilio G (2003) Neurodegeneration in amyotrophic lateral sclerosis: the role of oxidative stress and altered homeostasis of metals. Brain Res Bull 61(4):365–374. https://doi.org/10.1016/s0361-9230(03)00179-5

Chandel NS (2021) Lipid metabolism. Cold Spring Harb Perspect Biol 13(9):a040576. https://doi.org/10.1101/cshperspect.a040576

Chaves-Filho AB, Pinto IFD, Dantas LS, Xavier AM, Inague A, Faria RL et al (2019) Alterations in lipid metabolism of spinal cord linked to amyotrophic lateral sclerosis. Sci Rep 9(1):11642. https://doi.org/10.1038/s41598-019-48059-7

Chen B, Retzlaff M, Roos T, Frydman J (2011) Cellular strategies of protein quality control. Cold Spring Harb Perspect Biol 3(8):a004374. https://doi.org/10.1101/cshperspect.a004374

Chen S, Zhang X, Song L, Le W (2012) Autophagy dysregulation in amyotrophic lateral sclerosis: autophagy dysregulation in ALS. Brain Pathol 22(1):110–116. https://doi.org/10.1111/j.1750-3639.2011.00546.x

Chen L-X, Xu H-F, Wang P-S, Yang X-X, Wu Z-Y, Li H-F (2021) SOD1 mutation spectrum and natural history of ALS patients in a 15-year cohort in southeastern China. Front Genet 12:746060. https://doi.org/10.3389/fgene.2021.746060

Cheng H-C, Qi RZ, Paudel H, Zhu H-J (2011) Regulation and function of protein kinases and phosphatases. Enzyme Res 2011:1–3. https://doi.org/10.4061/2011/794089

Chua JP, De Calbiac H, Kabashi E, Barmada SJ (2022) Autophagy and ALS: mechanistic insights and therapeutic implications. Autophagy 18(2):254–282. https://doi.org/10.1080/15548627.2021.1926656

Coleman MP (2022) Axon biology in ALS: mechanisms of axon degeneration and prospects for therapy. Neurotherapeutics 19(4):1133–1144. https://doi.org/10.1007/s13311-022-01297-6

Consoli V, Sorrenti V, Grosso S, Vanella L (2021) Heme oxygenase-1 signaling and redox homeostasis in physiopathological conditions. Biomolecules 11(4):589. https://doi.org/10.3390/biom11040589

Cunha-Oliveira T, Montezinho L, Mendes C, Firuzi O, Saso L, Oliveira PJ et al (2020) Oxidative stress in amyotrophic lateral sclerosis: pathophysiology and opportunities for pharmacological intervention. Oxid Med Cell Longev 2020:1–29. https://doi.org/10.1155/2020/5021694

Espinosa-Diez C, Miguel V, Mennerich D, Kietzmann T, Sánchez-Pérez P, Cadenas S et al (2015) Antioxidant responses and cellular adjustments to oxidative stress. Redox Biol 6:183–197. https://doi.org/10.1016/j.redox.2015.07.008

Evans MC, Couch Y, Sibson N, Turner MR (2013) Inflammation and neurovascular changes in amyotrophic lateral sclerosis. Mol Cell Neurosci 53:34–41. https://doi.org/10.1016/j.mcn.2012.10.008

Fischer LR, Glass JD (2007) Axonal degeneration in motor neuron disease. Neurodegener Dis 4(6):431–442. https://doi.org/10.1159/000107704

Fröhlich F, Montoro AG (2023) The role of lysosomes in lipid homeostasis. Biol Chem 404(5):455–465. https://doi.org/10.1515/hsz-2022-0287

Fukano K, Kimura K (2014) Measurement of enolase activity in cell lysates. Methods Enzymol 542:115–124. https://doi.org/10.1016/B978-0-12-416618-9.00006-6

Gagliardi D, Ripellino P, Meneri M, Del Bo R, Antognozzi S, Comi GP et al (2023) Clinical and molecular features of patients with amyotrophic lateral sclerosis and SOD1 mutations: a monocentric study. Front Neurol 14:1169689. https://doi.org/10.3389/fneur.2023.1169689

Gandhi S, Abramov AY (2012) Mechanism of oxidative stress in neurodegeneration. Oxid Med Cell Longev 2012:1–11. https://doi.org/10.1155/2012/428010

Garbuzova-Davis S, Haller E, Saporta S, Kolomey I, Nicosia SV, Sanberg PR (2007) Ultrastructure of blood–brain barrier and blood–spinal cord barrier in SOD1 mice modeling ALS. Brain Res 1157:126–137. https://doi.org/10.1016/j.brainres.2007.04.044

Garbuzova-Davis S, Hernandez-Ontiveros DG, Rodrigues MCO, Haller E, Frisina-Deyo A, Mirtyl S et al (2012) Impaired blood–brain/spinal cord barrier in ALS patients. Brain Res 1469:114–128. https://doi.org/10.1016/j.brainres.2012.05.056

George DE, Tepe JJ (2021) Advances in proteasome enhancement by small molecules. Biomolecules 11(12):1789. https://doi.org/10.3390/biom11121789

Guo C, Sun L, Chen X, Zhang D (2013) Oxidative stress, mitochondrial damage and neurodegenerative diseases. Neural Regen Res 8(21):2003–2014. https://doi.org/10.3969/j.issn.1673-5374.2013.21.009

Han JH, Lee E-J, Park W, Ha K-T, Chung H-S (2023) Natural compounds as lactate dehydrogenase inhibitors: potential therapeutics for lactate dehydrogenase inhibitors-related diseases. Front Pharmacol 14:1275000. https://doi.org/10.3389/fphar.2023.1275000

Haouari S, Vourc'h P, Jeanne M, Marouillat S, Veyrat-Durebex C, Lanznaster D et al (2022) The roles of NEDD4 subfamily of HECT E3 ubiquitin ligases in neurodevelopment and neurodegeneration. Int J Mol Sci 23(7):3882. https://doi.org/10.3390/ijms23073882

Hosaka T, Tsuji H, Tamaoka A (2021) Biomolecular modifications linked to oxidative stress in amyotrophic lateral sclerosis: determining promising biomarkers related to oxidative stress. Processes (Basel) 9(9):1667. https://doi.org/10.3390/pr9091667

Hough MA, Grossmann JG, Antonyuk SV, Strange RW, Doucette PA, Rodriguez JA et al (2004) Dimer destabilization in superoxide dismutase may result in disease-causing properties: structures of motor neuron disease mutants. Proc Natl Acad Sci U S A 101(16):5976–5981. https://doi.org/10.1073/pnas.0305143101

Hu JH, Zhang H, Wagey R, Krieger C, Pelech SL (2003) Protein kinase and protein phosphatase expression in amyotrophic lateral sclerosis spinal cord. J Neurochem 85(2):432–442. https://doi.org/10.1046/j.1471-4159.2003.01670.x

Huang Q, Figueiredo-Pereira ME (2010) Ubiquitin/proteasome pathway impairment in neurodegeneration: therapeutic implications. Apoptosis 15(11):1292–1311. https://doi.org/10.1007/s10495-010-0466-z

Huijsman E, van de Par C, Economou C, van der Poel C, Lynch GS, Schoiswohl G et al (2009) Adipose triacylglycerol lipase deletion alters whole body energy metabolism and impairs exercise performance in mice. Am J Physiol Endocrinol Metab 297(2):E505–E513. https://doi.org/10.1152/ajpendo.00190.2009

Iacobazzi V, Infantino V (2014) Citrate—new functions for an old metabolite. Biol Chem 395(4):387–399. https://doi.org/10.1515/hsz-2013-0271

Israelsen WJ, Heiden MGV (2015) Pyruvate kinase: function, regulation and role in cancer. Semin Cell Dev Biol 43:43–51. https://doi.org/10.1016/j.semcdb.2015.08.004

Iwashima T, Tateishi T, Yamasaki R, Motomura K, Ohyagi Y, Kira J-I (2010) Two cases of familial amyotrophic lateral sclerosis with a SOD1L126S mutation showing high age at onset and slow progression. Rinsho Shinkeigaku 50(3):163–167. https://doi.org/10.5692/clinicalneurol.50.163

Jensen-Urstad APL, Semenkovich CF (2012) Fatty acid synthase and liver triglyceride metabolism: housekeeper or messenger? Biochim Biophys Acta 1821(5):747–753. https://doi.org/10.1016/j.bbalip.2011.09.017

Kametani F, Obi T, Shishido T, Akatsu H, Murayama S, Saito Y et al (2016) Mass spectrometric analysis of accumulated TDP-43 in amyotrophic lateral sclerosis brains. Sci Rep 6(1):23281. https://doi.org/10.1038/srep23281

Kawamata H, Manfredi G (2010) Mitochondrial dysfunction and intracellular calcium dysregulation in ALS. Mech Ageing Dev 131(7–8):517–526. https://doi.org/10.1016/j.mad.2010.05.003

Kim M-J, Bae J-H, Kim J-M, Kim HR, Yoon B-N, Sung J-J et al (2016) Rapid progression of sporadic ALS in a patient carrying SOD1 p.Gly13Arg mutation. Exp Neurobiol 25(6):347–350. https://doi.org/10.5607/en.2016.25.6.347

Lambert-Smith IA, Saunders DN, Yerbury JJ (2022) Proteostasis impairment and ALS. Prog Biophys Mol Biol 174:3–27. https://doi.org/10.1016/j.pbiomolbio.2022.06.001

Lee JK, Shin JH, Lee JE, Choi E-J (2015) Role of autophagy in the pathogenesis of amyotrophic lateral sclerosis. Biochim Biophys Acta 1852(11):2517–2524. https://doi.org/10.1016/j.bbadis.2015.08.005

Lee W-S, Ham W, Kim J (2021) Roles of NAD(P)H:Quinone oxidoreductase 1 in diverse diseases. Life (Basel) 11(12):1301. https://doi.org/10.3390/life11121301

Lee J, Lerskiatiphanich T, Marallag J (2022) Glucose metabolism in amyotrophic lateral sclerosis: it is bitter-sweet. Neural Regen Res 17(9):1975. https://doi.org/10.4103/1673-5374.335154

Levine B, Kroemer G (2008) Autophagy in the pathogenesis of disease. Cell 132(1):27–42. https://doi.org/10.1016/j.cell.2007.12.018

Lin H, Xu L, Yu S, Hong W, Huang M, Xu P (2020) Therapeutics targeting the fibrinolytic system. Exp Mol Med 52(3):367–379. https://doi.org/10.1038/s12276-020-0397-x

Lu C-H, Allen K, Oei F, Leoni E, Kuhle J, Tree T et al (2016) Systemic inflammatory response and neuromuscular involvement in amyotrophic lateral sclerosis. Neurol Neuroimmunol Neuroinflamm 3(4):e244. https://doi.org/10.1212/nxi.0000000000000244

Łukaszewicz-Zając M, Mroczko B, Słowik A (2014) Matrix metalloproteinases (MMPs) and their tissue inhibitors (TIMPs) in amyotrophic lateral sclerosis (ALS). J Neural Transm (Vienna) 121(11):1387–1397. https://doi.org/10.1007/s00702-014-1205-3

Maguire G (2017) Amyotrophic lateral sclerosis as a protein level, non-genomic disease: therapy with S2RM exosome released molecules. World J Stem Cells 9(11):187–202. https://doi.org/10.4252/wjsc.v9.i11.187

Malik AR, Willnow TE (2019) Excitatory amino acid transporters in physiology and disorders of the central nervous system. Int J Mol Sci 20(22):5671. https://doi.org/10.3390/ijms20225671

Martinez-Vicente M (2017) Neuronal mitophagy in neurodegenerative diseases. Front Mol Neurosci 10:64. https://doi.org/10.3389/fnmol.2017.00064

Monaco A, Fraldi A (2020) Protein aggregation and dysfunction of autophagy-lysosomal pathway: a vicious cycle in lysosomal storage diseases. Front Mol Neurosci 13:37. https://doi.org/10.3389/fnmol.2020.00037

Nelson AT, Trotti D (2022) Altered bioenergetics and metabolic homeostasis in amyotrophic lateral sclerosis. Neurotherapeutics 19(4):1102–1118. https://doi.org/10.1007/s13311-022-01262-3

Niego B, Medcalf RL (2014) Plasmin-dependent modulation of the blood-brain barrier: a major consideration during tPA-induced thrombolysis? J Cereb Blood Flow Metab 34(8):1283–1296. https://doi.org/10.1038/jcbfm.2014.99

Olivecrona G (2016) Role of lipoprotein lipase in lipid metabolism. Curr Opin Lipidol 27(3):233–241. https://doi.org/10.1097/MOL.0000000000000297

Pajarillo E, Rizor A, Lee J, Aschner M, Lee E (2019) The role of astrocytic glutamate transporters GLT-1 and GLAST in neurological disorders: potential targets for neurotherapeutics. Neuropharmacology 161:107559. https://doi.org/10.1016/j.neuropharm.2019.03.002

Park S, Jeon J-H, Min B-K, Ha C-M, Thoudam T, Park B-Y et al (2018) Role of the pyruvate dehydrogenase complex in metabolic remodeling: differential pyruvate dehydrogenase complex functions in metabolism. Diabetes Metab J 42(4):270. https://doi.org/10.4093/dmj.2018.0101

Parkin GM, Udawela M, Gibbons A, Dean B (2018) Glutamate transporters, EAAT1 and EAAT2, are potentially important in the pathophysiology and treatment of schizophrenia and affective disorders. World J Psychiatry 8(2):51–63. https://doi.org/10.5498/wjp.v8.i2.51

Peggion C, Scalcon V, Massimino ML, Nies K, Lopreiato R, Rigobello MP et al (2022) SOD1 in ALS: taking stock in pathogenic mechanisms and the role of glial and muscle cells. Antioxidants (Basel) 11(4):614. https://doi.org/10.3390/antiox11040614

Petrucelli L, Dawson T (2004) Mechanism of neurodegenerative disease: role of the ubiquitin proteasome system. Ann Med 36(4):315–320. https://doi.org/10.1080/07853890410031948

Prabhulkar S, Tian H, Wang X, Zhu J-J, Li C-Z (2012) Engineered proteins: redox properties and their applications. Antioxid Redox Signal 17(12):1796–1822. https://doi.org/10.1089/ars.2011.4001

Pralhada Rao R, Vaidyanathan N, Rengasamy M, Mammen Oommen A, Somaiya N, Jagannath MR (2013) Sphingolipid metabolic pathway: an overview of major roles played in human diseases. J Lipids 2013:1–12. https://doi.org/10.1155/2013/178910

Qian L, Chai AB, Gelissen IC, Brown AJ (2022) Balancing cholesterol in the brain: from synthesis to disposal. Explor Neuroprot Ther 2:1–27. https://doi.org/10.37349/ent.2022.00015

Quinet G, Gonzalez-Santamarta M, Louche C, Rodriguez MS (2020) Mechanisms regulating the UPS-ALS crosstalk: the role of proteaphagy. Molecules 25(10):2352. https://doi.org/10.3390/molecules25102352

Raghu P, Joseph A, Krishnan H, Singh P, Saha S (2019) Phosphoinositides: regulators of nervous system function in health and disease. Front Mol Neurosci 12:208. https://doi.org/10.3389/fnmol.2019.00208

Rakhit R, Chakrabartty A (2006) Structure, folding, and misfolding of Cu, Zn superoxide dismutase in amyotrophic lateral sclerosis. Biochim Biophys Acta 1762(11–12):1025–1037. https://doi.org/10.1016/j.bbadis.2006.05.004

Rempe RG, Hartz AMS, Bauer B (2016) Matrix metalloproteinases in the brain and blood-brain barrier: versatile breakers and makers. J Cereb Blood Flow Metab 36(9):1481–1507. https://doi.org/10.1177/0271678x16655551

Roberts DJ, Miyamoto S (2015) Hexokinase II integrates energy metabolism and cellular protection: akting on mitochondria and TORCing to autophagy. Cell Death Differ 22(2):248–257. https://doi.org/10.1038/cdd.2014.173

Ronowska A, Szutowicz A, Bielarczyk H, Gul-Hinc S, Klimaszewska-Łata J, Dyś A et al (2018) The regulatory effects of acetyl-CoA distribution in the healthy and diseased brain. Front Cell Neurosci 12:169. https://doi.org/10.3389/fncel.2018.00169

Rossi FH, Franco MC, Estevez AG (2013) Pathophysiology of amyotrophic lateral sclerosis. In: Current advances in amyotrophic lateral sclerosis. IntechOpen, London. https://doi.org/10.5772/56562

Rothstein JD, Tsai G, Kuncl RW, Clawson L, Cornblath DR, Drachman DB et al (1990) Abnormal excitatory amino acid metabolism in amyotrophic lateral sclerosis. Ann Neurol 28(1):18–25. https://doi.org/10.1002/ana.410280106

Ruegsegger C, Saxena S (2016) Proteostasis impairment in ALS. Brain Res 1648:571–579. https://doi.org/10.1016/j.brainres.2016.03.032

Saccon RA, Bunton-Stasyshyn RKA, Fisher EMC, Fratta P (2013) Is SOD1 loss of function involved in amyotrophic lateral sclerosis? Brain 136(8):2342–2358. https://doi.org/10.1093/brain/awt097

Sadler GL, Lewis KN, Narayana VK, De Souza DP, Mason J, McLean C et al (2022) Lipid metabolism is dysregulated in the motor cortex white matter in amyotrophic lateral sclerosis. Metabolites 12(6):554. https://doi.org/10.3390/metabo12060554

Schmidt MF, Gan ZY, Komander D, Dewson G (2021) Ubiquitin signalling in neurodegeneration: mechanisms and therapeutic opportunities. Cell Death Differ 28(2):570–590. https://doi.org/10.1038/s41418-020-00706-7

Sebastián D, Guitart M, García-Martínez C, Mauvezin C, Orellana-Gavaldà JM, Serra D et al (2009) Novel role of FATP1 in mitochondrial fatty acid oxidation in skeletal muscle cells. J Lipid Res 50(9):1789–1799. https://doi.org/10.1194/jlr.m800535-jlr200

Shukla V, Mishra SK, Pant HC (2011) Oxidative stress in neurodegeneration. Adv Pharmacol Sci 2011:1–13. https://doi.org/10.1155/2011/572634

Siddique N, Siddique T (2008) Genetics of amyotrophic lateral sclerosis. Phys Med Rehabil Clin N Am 19(3):429–439. https://doi.org/10.1016/j.pmr.2008.05.001

Stanton RC (2012) Glucose-6-phosphate dehydrogenase, NADPH, and cell survival. IUBMB Life 64(5):362–369. https://doi.org/10.1002/iub.1017

Stifanese R, Averna M, De Tullio R, Pedrazzi M, Milanese M, Bonifacino T et al (2014) Role of calpain-1 in the early phase of experimental ALS. Arch Biochem Biophys 562:1–8. https://doi.org/10.1016/j.abb.2014.08.006

Szelechowski M, Amoedo N, Obre E, Léger C, Allard L, Bonneu M et al (2018) Metabolic reprogramming in amyotrophic lateral sclerosis. Sci Rep 8(1):3953. https://doi.org/10.1038/s41598-018-22318-5

Tefera TW, Steyn FJ, Ngo ST, Borges K (2021) CNS glucose metabolism in amyotrophic lateral sclerosis: a therapeutic target? Cell Biosci 11(1):14. https://doi.org/10.1186/s13578-020-00511-2

Thomas KA, Rubin BH, Bier CJ, Richardson JS, Richardson DC (1974) The crystal structure of bovine Cu2+, Zn2+ superoxide dismutase at 5.5-A resolution. J Biol Chem 249(17):5677–5683. https://pubmed.ncbi.nlm.nih.gov/4472353/

Tommasini-Ghelfi S, Murnan K, Kouri FM, Mahajan AS, May JL, Stegh AH (2019) Cancer-associated mutation and beyond: the emerging biology of isocitrate dehydrogenases in human disease. Sci Adv 5(5):eaaw4543. https://doi.org/10.1126/sciadv.aaw4543

Trabjerg MS, Andersen DC, Huntjens P, Oklinski KE, Bolther L, Hald JL et al (2021) Downregulating carnitine palmitoyl transferase 1 affects disease progression in the SOD1 G93A mouse model of ALS. Commun Biol 4(1):509. https://doi.org/10.1038/s42003-021-02034-z

Trostchansky A (2019) Overview of lipid biomarkers in amyotrophic lateral sclerosis (ALS). In: Advances in experimental medicine and biology. Springer International Publishing, Cham, pp 233–241. https://doi.org/10.1007/978-3-030-21735-8_18

Veith A, Moorthy B (2018) Role of cytochrome P450s in the generation and metabolism of reactive oxygen species. Curr Opin Toxicol 7:44–51. https://doi.org/10.1016/j.cotox.2017.10.003

Vosler PS, Brennan CS, Chen J (2008) Calpain-mediated signaling mechanisms in neuronal injury and neurodegeneration. Mol Neurobiol 38(1):78–100. https://doi.org/10.1007/s12035-008-8036-x

Walker AK (2010) Protein disulfide isomerase and the endoplasmic reticulum in amyotrophic lateral sclerosis. J Neurosci 30(11):3865–3867. https://doi.org/10.1523/jneurosci.0408-10.2010

Wang T, Wang J, Hu X, Huang X, Chen G-X (2020) Current understanding of glucose transporter 4 expression and functional mechanisms. World J Biol Chem 11(3):76–98. https://doi.org/10.4331/wjbc.v11.i3.76

Wootz H, Weber E, Korhonen L, Lindholm D (2006) Altered distribution and levels of cathepsin D and cystatins in amyotrophic lateral sclerosis transgenic mice: possible roles in motor neuron survival. Neuroscience 143(2):419–430. https://doi.org/10.1016/j.neuroscience.2006.07.048

Wright GSA, Antonyuk SV, Hasnain SS (2019) The biophysics of superoxide dismutase-1 and amyotrophic lateral sclerosis. Q Rev Biophys 52:e12. https://doi.org/10.1017/s003358351900012x

Yang H, Wu Y, Yen W, Liu H, Hwang T, Stern A et al (2019) The redox role of G6PD in cell growth, cell death, and cancer. Cells 8(9):1055. https://doi.org/10.3390/cells8091055

Yepes M, Woo Y, Martin-Jimenez C (2021) Plasminogen activators in neurovascular and neurodegenerative disorders. Int J Mol Sci 22(9):4380. https://doi.org/10.3390/ijms22094380

Yi M, Ban Y, Tan Y, Xiong W, Li G, Xiang B (2019) 6-Phosphofructo-2-kinase/fructose-2,6-biphosphatase 3 and 4: a pair of valves for fine-tuning of glucose metabolism in human cancer. Mol Metab 20:1–13. https://doi.org/10.1016/j.molmet.2018.11.013

Zhang Y, Roh YJ, Han S-J, Park I, Lee HM, Ok YS et al (2020) Role of selenoproteins in redox regulation of signaling and the antioxidant system: a review. Antioxidants (Basel) 9(5):383. https://doi.org/10.3390/antiox9050383

Enzymes Involved in Multiple Sclerosis 4

Jobin Jose and Priyam Purkait

Abstract

Multiple sclerosis (MS) is a long-term chronic disease that affects the central nervous system, the network of nerves responsible for controlling most body functions. Millions of people worldwide are affected by MS. The disease disrupts communication between the brain and body, leading to a variety of symptoms. The exact cause of MS remains unknown, but scientific evidence points toward a malfunction of the immune system. Usually, the immune system defends the body against germs. In MS, however, the immune system mistakenly attacks the myelin sheath, a fatty substance that insulates and protects nerve fibers. This damage disrupts nerve signals, leading to the characteristic symptoms of MS. Immune cells, like other cells in the body, require a steady supply of amino acids to function correctly. When the immune system is activated due to inflammation or foreign invaders, its demand for amino acids increases rapidly. Studies suggested that changes in amino acid levels can influence immune cell activity and potentially contribute to MS development. In MS, oxidative stress encourages tissue damage. Antioxidant enzymes such as glutathione peroxidase-4, heme oxygenase-1, peroxiredoxin-5, and superoxide dismutase-1 may be increased to guard against this. This chapter mainly deals with the role of various enzymes involved in MS.

Keywords

Multiple sclerosis · Myelin sheath · Reactive oxygen species · Metabolic pathways

J. Jose (✉) · P. Purkait
Department of Pharmaceutics, NITTE Deemed-to-be University, NGSM Institute of Pharmaceutical Sciences, Mangalore, Karnataka, India

B. Mathew, D. G. T. Parambi (eds.), *Enzymes in Neurodegenerative Disorders*,
https://doi.org/10.1007/978-981-97-6822-6_4

4.1 Introduction

Multiple sclerosis (MS) is a long-term (chronic) condition that causes inflammation and damage to the central nervous system (CNS). The exact cause of MS is unknown, but it is likely a combination of factors (multifactorial) (Schreibelt et al. 2007; Correale 2021). MS was thought to be mediated by T cells attacking the myelin sheath (Dobson and Giovannoni 2019). MS damages myelin, the protective sheath around nerve fibers, in the CNS. This disrupts nerve signals and can lead to a variety of symptoms. MS affects people differently, with some experiencing flare-ups followed by periods of stability, while others gradually worsen symptoms (Goldenberg 2012).

How our bodies break down amino acids (AAs) significantly impacts our immune system, but the effects can be surprisingly diverse and contradictory. Researchers are untangling the intricate mechanisms by which these breakdown pathways, particularly those involving tryptophan (Trp) and arginine (Arg), influence immune responses. The key pathways are mediated by enzymes like indoleamine 2,3-dioxygenase 1 (IDO1) and ARG1, highlighting their role in the autoimmune response and how they can have both cytotoxic and neuroprotective effects. Trp and Arg metabolisms act as critical control points within the immune system, offering potential targets for therapeutic intervention. These pathways are relevant in various conditions, including autoimmune diseases like MS, tumors, and situations like organ transplants where the body might reject the foreign tissue. As scientists discover new components within these pathways, the interplay between factors that either stimulate or suppress the immune system becomes even more fascinating. Despite ongoing research, there are still uncertainties regarding the exact role these pathways play in immune regulation, neuroprotection, and excitotoxicity (excessive stimulation of neurons leading to cell death). However, the potential to manipulate these biochemical pathways for treating various diseases is an up-and-coming area of exploration (Correale 2021; Goldenberg 2012). This chapter mainly discusses the role of different enzymes involved in MS.

Proteases, also known as proteinases or peptidases, are enzymes that play a crucial role in our bodies. They act like molecular scissors, cutting proteins at specific points by breaking down the chemical bonds that hold AAs together. Hundreds of these proteases exist in the human body, encoded by our genes, and they make up a significant portion of our genetic blueprint. While scientists link proteases to various diseases, the exact function of many remains unclear. This includes what proteins they target (substrates), how they are activated, and how they are naturally controlled (inhibited). Understanding their role within the body is complex because proteases often work together in intricate pathways, interacting with other enzymes, target proteins, binding molecules, and inhibitors within the cellular environment. Proteases are categorized into different classes based on their chemical structure and how they cut protein bonds (Scarisbrick 2008).

Studies on the enzymes called matrix metalloproteinases (MMPs) and their inhibitors in MS brain lesions and fluids, along with immune cell activity, suggested that MMPs play a role in the development of MS. Additionally, differences in MMP levels in body fluids may help identify MS types and disease activity (Boziki and Grigoriadis 2018). Serine proteases are not only known for their roles in blood clotting and breakdown but also play essential functions in the nervous system. These enzymes are involved in various processes, including cell migration, nerve fiber growth, and guidance, remodeling of synapses (connections between nerve cells), regulation of nerve cell activity, survival of both glial cells (supportive cells in the nervous system), and neurons (nerve cells). Researchers have identified a specific serine protease called myelencephalon-specific protease (MSP). This enzyme is mainly found in the adult CNS and is present in neurons and oligodendroglia (cells that produce myelin). Scientists investigated the potential role of MSP in demyelination, a process where the protective fatty sheath (myelin) surrounding nerve fibers is damaged. They examined MSP levels in MS lesions and animal disease models. The findings revealed high levels of MSP within immune cells that accumulate at demyelination sites in both MS lesions and animal models. This suggests that excess MSP, present during inflammatory conditions in the CNS, may contribute to demyelination (Scarisbrick et al. 2002).

Many studies investigated how glutamate levels are altered in MS. Glutamate is a signaling molecule in the nervous system, but excess glutamate can be toxic. One factor controlling glutamate levels outside cells is a transporter called the cystine/glutamate antiporter, also known as system xc−. This transporter is made of two parts: xCT and 4F2hc. The xCT subunit determines which AAs are transported, while 4F2hc is needed for xCT to reach the cell surface and is involved in other transport systems. The cystine/glutamate antiporter imports cystine, a building block for the antioxidant glutathione. This study showed that activating human monocytes releases glutamate through system xc− and increases their production of the xCT subunit. They found that xCT expression is elevated in cells of the monocyte-macrophage-microglia lineage in both EAE (an animal model of MS) and MS itself. This was observed in the CNS and peripheral blood cells. Notably, these immune cells had higher xCT levels in MS and EAE than healthy controls. These findings suggest a connection between immune activation, increased xCT levels, and glutamate excitotoxicity, which may contribute to damage to oligodendrocytes (cells that produce myelin) in MS (Pampliega et al. 2011).

Glutamate and aspartate (L-aspartate) are the brain's primary excitatory AAs, playing a crucial role in development, communication between nerve cells, and learning. Interestingly, another form of aspartate called D-aspartate can also activate specific receptors involved in glutamate signaling. However, unlike glutamate, the link between D-aspartate, its mirror image L-aspartate, and MS progression with inflammation remains unclear. While the role of L-aspartate as a signaling molecule is debated, it is considered the second most important excitatory AA. Some evidence suggests it might be released alongside glutamate. Studies also showed that L-aspartate can directly activate specific glutamate receptors, suggesting changes in its metabolism within nerve cells or glial cells (supportive cells) might contribute to the development of dysfunctional synapses and, eventually, cell death (excitotoxicity) in MS (Errico et al. 2023).

Threonine, a naturally occurring AA, is used to treat spasticity in MS patients. Spasticity is a condition causing muscle stiffness and tightness. The researchers hypothesized that threonine might improve spasticity by enhancing a specific inhibitory pathway in the nervous system. Their findings showed a modest decrease in spasticity scores among treated patients compared to a control group. However, the study did not observe any improvement in overall symptoms or general neurological function. Despite the lack of broader benefits, the reduction in spasticity scores suggested that threonine might have therapeutic potential. Further research is needed to explore the use of higher doses or combinations with other treatments to determine if threonine can be a viable option for managing spasticity in MS (Hauser et al. 1992).

Gamma-aminobutyric acid (GABA) is a chemical messenger in the brain that acts like a brake on nerve cell activity. It is produced from glutamate, another important neurotransmitter, by an enzyme called glutamic acid decarboxylase. GABA plays a role in various brain functions like learning, memory, and movement coordination. Another enzyme, GABAT, breaks down GABA and eventually enters the energy production cycle within cells. Interestingly, GABA also seems to have anti-inflammatory properties. Researchers suggested that gut bacteria, the vast community of microbes living in our intestines, might influence MS. Some gut bacteria produce GABAT, the enzyme that degrades GABA. A high abundance of these bacteria could potentially lead to lower GABA levels in specific brain regions. This decrease in GABA, along with potentially lower glutamate levels, might contribute to MS development and the physical disability experienced by some patients (Hussain and Muccee 2024).

Regulatory T cells, also known as Treg cells, are a vital part of the immune system. They act like natural brakes, stopping the immune system from attacking healthy tissues. These cells express a protein called FOXP3, which is crucial for their function. Treg cells, mostly CD4+ T cells with a specific protein (CD25) on their surface, can calm overactive immune responses. They can inhibit the function of several immune cells, such as natural killer cells, T cells, and antigen-presenting cells. This ability makes them essential for preventing autoimmune diseases, allergies, and rejection of transplanted organs. However, there's a flip side. Treg cells can also suppress the immune system's attack on tumors. This highlights the complex role of Treg cells in maintaining immune balance. While earlier attempts to define Treg cells based on their molecules were unsuccessful, research in this area has intensified. Scientists are now looking at FOXP3+ Treg cells as a potential target for new treatments (Sakaguchi et al. 2010).

The body's kynurenine pathway (KP) breaks down Trp, an essential AA. This pathway is involved in various functions but can also be activated in diseases like MS (Schwarcz 2004). When the KP becomes too active, it can produce harmful molecules that damage brain cells (Stone and Darlington 2013). Monocytes, a type of immune cell, are particularly affected by high levels of inflammation. This inflammation triggers the KP in monocytes, causing them to produce and release more of these harmful molecules (Jones et al. 2015). While the exact cause of MS remains unknown, research suggests that the KP might play a role. When there's inflammation in MS, the KP seems to activate in monocytes, producing these damaging molecules. These molecules can then travel to the brain and contribute to MS by harming brain cells and hindering repair mechanisms. Studies on MS patients and animal models showed that the KP is activated in the early stages of the disease (Torok et al. 2016).

The OxPhos activity has been the main focus of most research, and many studies have shown significant abnormalities in OxPhos activity in MS. The expression of tricarboxylic acid (TCA) cycle flux and glycolysis-related enzymes in the MS tissue. The metabolic process known as glycolysis is where a variety of enzymes, notably the rate-limiting enzymes pyruvate kinase (PK) and hexokinase (HK), convert glucose into pyruvate (Dringen et al. 2005). Hexokinase 2 (HK2) has been demonstrated to be the primary regulated isoform among the three distinct isoforms of HK. HK2 quickly phosphorylates glucose to glucose-6-phosphate, the first step in glycolysis. PK catalyzes the last phase of glycolysis (Yang et al. 2012; Gibson et al. 2000). The pyruvate generated during glycolysis can be carried into the mitochondria, where pyruvate dehydrogenase transforms it into acetyl-CoA, which powers the TCA cycle. α-Ketoglutarate dehydrogenase and malate dehydrogenase, which both produce NADH, are two additional rate-limiting TCA cycle enzymes (Gietl 1992; Patel and Roche 1990; Pierrot-Deseilligny and Souberbielle 2010). The most significant factor influencing OxPhos is NADH, which is closely related to TCA cycle metabolism and OxPhos (Sawcer et al. 2011).

4.2 Endogenous Antioxidant Enzymes

Like other organs, the brain must maintain a healthy balance between the natural production of oxygen molecules and their potential for causing harm. To achieve this, the CNS has a built-in antioxidant defense system. This system contains special enzymes that can neutralize these harmful molecules, called reactive oxygen species (ROS) (Itoh et al. 2003). The brain uses a lot of oxygen and contains fats susceptible to damage by ROS. When ROS overwhelm the antioxidants, they can damage essential molecules like fats, proteins, and DNA in brain cells. This damage can be seen in the brain tissue and spinal fluid of MS patients. Scientists are still figuring out which cell types are most affected by ROS in MS lesions. The body naturally produces antioxidant enzymes to fight ROS, which are often increased in diseases like Alzheimer's and Parkinson's.

Interestingly, studies using an animal model of MS have shown that boosting these antioxidant enzymes can improve the condition. Recent proteomics research has revealed that several antioxidant enzymes are significantly increased in MS lesions. This suggests the body is trying to combat oxidative stress (Van Horssen et al. 2008). When ROS levels rise due to stress or other factors, a protein called Nrf2 becomes activated. Usually, Nrf2 is kept inactive by another protein (Keap1). However, when ROS levels increase, Nrf2 breaks free from Keap1 and travels to the cell's control center (nucleus) (Kensler et al. 2007). Once in the nucleus, Nrf2 acts like a switch, turning on genes that produce various antioxidant enzymes. Scientists have discovered a large number (over 200) of genes controlled by Nrf2. All these genes play a role in combating oxidative stress and protecting the brain from damage (McCord and Edeas 2005; Thimmulappa et al. 2002; Kim et al. 2007). Gene transcription of most antioxidant enzymes is regulated through the nuclear factor-E2-related factor (Nrf2) transcription factor and antioxidant response elements (ARE) in the genes encoding enzymatic antioxidants. Under physiological conditions, Nrf2 is linked to the actin-bound Kelch-like ECH-associated protein 1 (Keap1) and is located in the cytoplasm. However, upon oxidative stress, Nrf2 dissociates from Keap1 and translocates to the nucleus, where it activates ARE-mediated gene transcription and induces the coordinate transcription of ARE-regulated genes. To date, over 200 Nrf2–ARE-driven genes involved in detoxification and antioxidant defense have been identified, including superoxide dismutases (SODs), peroxiredoxins (Prxs), heme oxygenases (HOs), glutathione peroxidases (GPxs), catalase, NAD(P)H:quinone oxidoreductase 1 (NQO1), and NHR:quinone oxidoreductase 2 (NQO2).

4.2.1 Superoxide Dismutases

Superoxide dismutase (SOD) enzymes are a crucial defense system within cells, particularly abundant in the CNS (Johnson and Giulivi 2005). They act as antioxidants, explicitly targeting a harmful molecule called superoxide anion. Superoxide is a byproduct of normal cellular processes, but it can damage cells at high levels. There are three main types of SOD found in humans, each with a different metal atom at its active site and varying locations within the cell: Cu/ZnSOD (superoxide dismutase-1 [SOD1]), found mainly in the cytoplasm of astrocytes and to a lesser extent in neurons; MnSOD (SOD2), primarily located within mitochondria, especially in neurons; and extracellular Cu/ZnSOD (SOD3), present outside cells but at lower levels than SOD1 and SOD2. Increased SOD expression has been observed in various diseases associated with oxidative stress, such as Alzheimer's disease and stroke.

Interestingly, some forms of amyotrophic lateral sclerosis are linked to mutations in the SOD1 gene, suggesting a complex role for these enzymes in neurodegeneration. Studies on MS revealed elevated SOD1 gene expression in active lesions, suggesting a potential involvement of this enzyme. However, attempts to directly administer SOD as a treatment for MS in animal models have not been successful. Recent research has developed synthetic SOD mimics that show promise in treating other neurological conditions. Investigating the effects of these mimics on MS could be a valuable area for future exploration. Additionally, combining SOD therapy with enzymes that remove hydrogen peroxide (a product of SOD activity) might be a more effective strategy for managing neuroinflammatory diseases (Marklund 1982).

4.2.2 Catalase

Catalase is a crucial antioxidant enzyme found within cells, particularly in structures called peroxisomes. It is critical in protecting cells from oxidative stress caused by hydrogen peroxide, a harmful molecule (Dringen et al. 2005). Catalase is a complex enzyme made of four identical subunits. Each subunit contains a unique molecule called heme, similar to what is found in the red blood cells, and NADPH, a cellular energy carrier. Depending on the amount of hydrogen peroxide present, catalase can act in two ways:

High hydrogen peroxide levels: When there is a lot of hydrogen peroxide, catalase efficiently breaks it down into harmless water and oxygen molecules. Limited glutathione availability: When another antioxidant called glutathione is scarce, catalase becomes even more critical in managing oxidative stress. All cell types within the CNS naturally produce catalase. Studies on animals with a condition similar to MS (EAE) have shown decreased catalase activity and gene expression, suggesting a potential link between reduced catalase and neuroinflammation. The studies suggested that increasing catalase levels might be beneficial in managing neuroinflammatory conditions. Studies using various methods to deliver catalase to animals with EAE have shown promising results: reduced demyelination (damage to the fatty nerve sheath), improved blood-brain barrier integrity, lessened neurological symptoms, and milder disease progression. These findings suggested that catalase could be a potential therapeutic strategy for MS, although further research is needed (Singh et al. 2004).

4.2.3 GPxs

GPx is a group of enzymes containing selenium that helps cells eliminate harmful peroxides and hydrogen peroxide. These molecules can damage cellular components if not properly neutralized. Mammals have six known types of GPx, each found in specific locations within cells or tissues. In the brain, GPx1, a particular type of GPx, is even more active than catalase, another antioxidant enzyme. This is particularly important because GPx1 is located in the cytosol and mitochondria, where a lot of superoxide radicals (precursors to hydrogen peroxide) are produced. This suggests that GPx might be a more critical defense system against oxidative stress in the CNS than catalase, mainly found in peroxisomes. Studies have shown increased GPx gene expression in areas of active demyelination in MS patients.

Additionally, research using an animal model of MS (EAE) suggested that GPx treatment might be beneficial. Treatment with GPx helped preserve the integrity of the blood-brain barrier, a protective layer around the brain. These findings highlighted the potential role of GPx in managing neuroinflammatory diseases like MS (Herbette et al. 2007).

4.2.4 Peroxiredoxins

Prxs are a recently discovered family of enzymes that have emerged as potentially significant players in eliminating hydrogen peroxide within mammalian cells. These versatile proteins go beyond just hydrogen peroxide removal. They can also break down other harmful peroxides and molecules like peroxynitrite. Additionally, Prxs play a role in regulating cell signaling, proliferation, differentiation, and gene expression. Six known groups of Prxs are classified based on the number and arrangement of specific AAs called cysteines in their active sites. These enzymes are found throughout the cell in various locations, including the cytosol, mitochondria, peroxisomes, and even outside the cell. Interestingly, Prx1 is concentrated in astrocytes in the brain, while Prx2 is more prevalent in neurons. Studies on other neurodegenerative diseases linked to oxidative stress suggest that Prx expression varies depending on the condition. For example, some Prx types were found to be increased, while others were decreased in patients with Alzheimer's disease. This highlights the complex role of Prxs in these diseases. While the involvement of Prxs has been explored in other neurological conditions, their role in MS remains unknown. Their potential as a critical antioxidant defense system warrants further investigation to understand their contribution to MS development (Sarafian et al. 1999).

4.2.5 HO

HO is an enzyme critical for heme breakdown, a natural process in the body. There are two main types: HO-1 (inducible): This type can be ramped up in response to various stimuli, including stress and its substrate, heme. It has antioxidant and anti-inflammatory properties. HO-2 (constitutive): This type is constantly present and plays a role in regular heme metabolism. Research on HO's role in MS has yielded mixed results. Some studies suggest that upregulating HO-1 might be beneficial: Experiments using EAE (a model of MS) showed that increasing HO-1 activity lessened disease severity. Inhibiting HO-1 might also be helpful: Other studies using EAE found that blocking HO-1 activity improved outcomes. These conflicting findings could be due to differences in the types of HO inhibitors used or variations between species. Studies have observed increased HO-1 activity in specific cell types within MS lesions, including hypertrophic astrocytes (enlarged glial cells), reactive microglia (immune cells), infiltrated macrophages (immune cells), and potential benefits of HO breakdown products. Heme breakdown by HO generates several products, some of which might be useful in MS treatment:

Biliverdin: This molecule is rapidly converted to bilirubin, a potent antioxidant. Studies suggest bilirubin can lessen disease progression and protect nerve cells.

Bilirubin: Despite its potential benefits, bilirubin can also be toxic. However, research shows it can still protect nerve cells from damage.

Findings suggest that HO, or the molecules it produces, could be promising targets for MS therapies. Further research is needed to understand the complex role of HO in MS and develop effective treatment strategies (Emerson and LeVine 2000; Liu et al. 2006).

4.2.6 Quinone Oxidoreductases

NQO1 and NQO2 are two enzymes found within cells (cytosolic) that contain a molecule called flavin. These enzymes play a dual role:

Neutralizing Harmful Molecules: They can inactivate quinones and related compounds, preventing them from contributing to cellular damage.

Broad-Spectrum Antioxidant Activity: They also possess general antioxidant properties, helping protect cells from oxidative stress. NQO1 is especially crucial in tissues with high antioxidant needs, such as the lungs and the CNS. In the healthy brain, NQO1 is primarily located in astrocytes (glial cells) and endothelial cells (lining blood vessels). Interestingly, increased NQO1 activity has been observed in astrocytes of patients with Alzheimer's and Parkinson's diseases, suggesting a potential response to oxidative stress in these conditions. Recent research has revealed a significant increase in NQO1 levels within inflammatory lesions in MS patients. These elevated levels were mainly found in hypertrophic astrocytes (enlarged glial cells) and macrophages containing myelin debris. Unlike NQO1, NQO2 is found at deficient levels in the brain. While it is expressed in other organs like the kidney, liver, and heart, its role in neurodegenerative or neuroinflammatory diseases remains largely unknown. Further research is needed to understand how NQO2 might be involved in these conditions (Long and Jaiswal 2000).

ROS produced by macrophages might trigger this increase and could act as a protective mechanism against cellular damage. Another antioxidant enzyme, PRDX2, is also upregulated in white matter MS lesions, mainly in astrocytes. The level of expression seems to correlate with the degree of inflammation and oxidative stress. This suggests that PRDX2 expression helps protect astrocytes from oxidative damage.

4.3 Impact of Arg Metabolism

Arg is an AA essential for mammals under specific conditions, like during an immune response. Its breakdown pathway is crucial in regulating the immune system, mainly through the enzyme arginase-1 (ARG1). ARG1 activity is essential in M2 macrophages, a type of immune cell associated with tissue repair and anti-inflammatory responses. These cells express ARG1 in response to various signaling molecules, including interleukin-4 (IL-4), IL-6, IL-13, and transforming growth factor-beta (TGF-β). In humans, unlike mice, ARG1 is present even without stimulation in specific structures within neutrophils (another immune cell type). ARG1 activity can limit the effectiveness of the immune response by depleting Arg, which is crucial for T-cell activation and proliferation. This depletion essentially halts T-cell activity in a specific growth phase. Recent research suggested a link between ARG1 and another immune regulatory enzyme, IDO1. In the presence of TGF-β, these enzymes can work together in immune regulation within dendritic cells (DCs), another key player in the immune system. Notably, ARG1 activity seems to be a prerequisite for IDO1 upregulation, suggesting a potential sequence of events in this regulatory pathway. The breakdown products of Arg, particularly polyamines like spermidine, also play a role in immune regulation. Spermidine can even mimic the effect of TGF-β in activating a signaling pathway that influences IDO1 activity within DCs. This complex interplay between ARG1, IDO1, and their metabolites highlights a potential mechanism for maintaining long-term immune tolerance (Correale 2021).

4.4 Conclusion

Future treatment strategies for MS might involve fine-tuning catabolism. In some cases, inhibiting the breakdown of specific AAs could enhance immunity; in others, increasing their breakdown might be beneficial in suppressing an overactive immune response. The potential therapeutic use of metabolites derived from Trp breakdown that have immunosuppressive properties is also under investigation. ROS are highly reactive molecules contributing to tissue damage to MS lesions. Studies suggest the nervous system attempts to compensate by increasing the expression of antioxidant enzymes regulated by the Nrf2/ARE pathway, as observed in animal MS (EAE) models and MS tissues. This highlights the potential of antioxidant therapy for MS. Delivering sufficient amounts of these molecules effectively to the CNS may be difficult. An alternative approach involves activating the Nrf2/ARE pathway within the body. This pathway can stimulate the production of the body's antioxidant enzymes, potentially achieving sufficient levels within the CNS to counteract the damaging effects of ROS in MS lesions. Further studies will determine the effectiveness of Nrf2/ARE activators in treating neuroinflammatory diseases like MS. Studies have observed increased expression of the antioxidant enzyme NQO1 in macrophages and astrocytes within MS lesions. The interaction between variables stimulating or repressing the immune system gets even more fascinating when scientists identify additional elements within these pathways.

References

Boziki M, Grigoriadis N (2018) An update on the role of matrix metalloproteinases in the pathogenesis of multiple sclerosis. Med Chem 14(2):155–169

Correale J (2021) Immunosuppressive amino-acid catabolizing enzymes in multiple sclerosis. Front Immunol 11:600428
Dobson R, Giovannoni G (2019) Multiple sclerosis—a review. Eur J Neurol 26(1):27–40
Dringen R, Pawlowski PG, Hirrlinger J (2005) Peroxide detoxification by brain cells. J Neurosci Res 79(1–2):157–165
Emerson MR, LeVine SM (2000) Heme oxygenase-1 and NADPH cytochrome P450 reductase expression in experimental allergic encephalomyelitis: an expanded view of the stress response. J Neurochem 75(6):2555–2562
Errico F, Gilio L, Mancini A, Nuzzo T, Bassi MS, Bellingacci L, Buttari F, Dolcetti E, Bruno A, Galifi G, Furlan R (2023) Cerebrospinal fluid, brain, and spinal cord levels of L-aspartate signal excitatory neurotransmission abnormalities in multiple sclerosis patients and experimental autoimmune encephalomyelitis mouse model. J Neurochem 166(3):534–546
Gibson GE, Park LCH, Sheu KFR, Blass JP, Calingasan NY (2000) The alpha-ketoglutarate dehydrogenase complex in neurodegeneration. Neurochem Int 36:97–112
Gietl C (1992) Malate dehydrogenase isoenzymes: cellular locations and role in the flow of metabolites between the cytoplasm and cell organelles. Biochim Biophys Acta 1100:217–234
Goldenberg MM (2012) Multiple sclerosis review. P T 37(3):175
Hauser SL, Doolittle TH, Lopez-Bresnahan M, Shahani B, Schoenfeld D, Shih VE, Growdon J, Lehrich JR (1992) An antispasticity effect of threonine in multiple sclerosis. Arch Neurol 49(9):923–926
Herbette S, Roeckel-Drevet P, Drevet JR (2007) Seleno-independent glutathione peroxidases: more than simple antioxidant scavengers. FEBS J 274(9):2163–2180
Hussain N, Muccee F (2024) In-silico characterization of GABAT protein found in gut-brain axis associated bacteria of healthy individuals and multiple sclerosis patients. Saudi J Biol Sci 31:103939
Itoh K, Wakabayashi N, Katoh Y, Ishii T, O'Connor T, Yamamoto M (2003) Keap1 regulates both cytoplasmic-nuclear shuttling and degradation of Nrf2 in response to electrophiles. Genes Cells 8(4):379–391
Johnson F, Giulivi C (2005) Superoxide dismutases and their impact upon human health. Mol Asp Med 26(4–5):340–352
Jones SP, Franco NF, Varney B, Sundaram G, Brown DA, De Bie J et al (2015) Expression of the kynurenine pathway in human peripheral blood mononuclear cells: implications for inflammatory and neurodegenerative disease. PLoS One 10(6):e0131389
Kensler TW, Wakabayashi N, Biswal S (2007) Cell survival responses to environmental stresses via the Keap1-Nrf2-ARE pathway. Annu Rev Pharmacol Toxicol 47:89–116
Kim YJ, Ahn JY, Liang P, Ip C, Zhang Y, Park YM (2007) Human prx1 gene is a target of Nrf2 and is up-regulated by hypoxia/reoxygenation: implication to tumor biology. Cancer Res 67(2):546–554
Liu Y, Liu J, Tetzlaff W, Paty DW, Cynader MS (2006) Biliverdin reductase, a major physiologic cytoprotectant, suppresses experimental autoimmune encephalomyelitis. Free Radic Biol Med 40(6):960–967
Long DJ II, Jaiswal AK (2000) NRH:quinone oxidoreductase2 (NQO2). Chem Biol Interact 129(1–2):99–112
Marklund SL (1982) Human copper-containing superoxide dismutase of high molecular weight. Proc Natl Acad Sci 79(24):7634–7638
McCord JM, Edeas MA (2005) SOD, oxidative stress and human pathologies: a brief history and a future vision. Biomed Pharmacother 59(4):139–142
Pampliega O, Domercq M, Soria FN, Villoslada P, Rodríguez-Antigüedad A, Matute C (2011) Increased expression of cystine/glutamate antiporter in multiple sclerosis. J Neuroinflammation 8:1–2
Patel MS, Roche TE (1990) Molecular biology and biochemistry of pyruvate dehydrogenase complexes. FASEB J 4:3224–3233
Pierrot-Deseilligny C, Souberbielle JC (2010) Is hypovitaminosis D one of the environmental risk factors for multiple sclerosis? Brain 133:1869–1888

Sakaguchi S, Miyara M, Costantino CM, Hafler DA (2010) FOXP3+ regulatory T cells in the human immune system. Nat Rev Immunol 10(7):490–500

Sarafian TA, Verity MA, Vinters HV, Shih CC, Shi L, Ji XD, Dong L, Shau H (1999) Differential expression of peroxiredoxin subtypes in human brain cell types. J Neurosci Res 56(2):206–212

Sawcer S, Hellenthal G, Pirinen M et al (2011) Genetic risk and a primary role for cell-mediated immune mechanisms in multiple sclerosis. Nature 476:214–219

Scarisbrick IA (2008) The multiple sclerosis degradome: enzymatic cascades in development and progression of central nervous system inflammatory disease. Curr Top Microbiol Immunol 318:133–175

Scarisbrick IA, Blaber SI, Lucchinetti CF, Genain CP, Blaber M, Rodriguez M (2002) Activity of a newly identified serine protease in CNS demyelination. Brain 125(6):1283–1296

Schreibelt G, Van Horssen J, Van Rossum S, Dijkstra CD, Drukarch B, de Vries HE (2007) Therapeutic potential and biological role of endogenous antioxidant enzymes in multiple sclerosis pathology. Brain Res Rev 56(2):322–330

Schwarcz R (2004) The kynurenine pathway of tryptophan degradation as a drug target. Curr Opin Pharmacol 4(1):12–17

Singh I, Paintlia AS, Khan M, Stanislaus R, Paintlia MK, Haq E, Singh AK, Contreras MA (2004) Impaired peroxisomal function in the central nervous system with inflammatory disease of experimental autoimmune encephalomyelitis animals and protection by lovastatin treatment. Brain Res 1022(1–2):1

Stone TW, Darlington LG (2013) The kynurenine pathway as a therapeutic target in cognitive and neurodegenerative disorders. Br J Pharmacol 169(6):1211–1227

Thimmulappa RK, Mai KH, Srisuma S, Kensler TW, Yamamoto M, Biswal S (2002) Identification of Nrf2-regulated genes induced by the chemopreventive agent sulforaphane by oligonucleotide microarray. Cancer Res 62(18):5196–5203

Torok N, Majlath Z, Fulop F, Toldi J, Vecsei L (2016) Brain aging and disorders of the central nervous system: kynurenines and drug metabolism. Curr Drug Metab 17(5):412–429

Van Horssen J, Schreibelt G, Drexhage J, Hazes T, Dijkstra CD, Van der Valk P, De Vries HE (2008) Severe oxidative damage in multiple sclerosis lesions coincides with enhanced antioxidant enzyme expression. Free Radic Biol Med 45(12):1729–1737

Yang W, Xia Y, Hawke D, Li X, Liang J, Xing D et al (2012) PKM2 phosphorylates histone H3 and promotes gene transcription and tumorigenesis. Cell 150:685–696

Role of Enzymes in Epilepsy

5

Manal Mohammed, P. Prabitha, and Tariq G. Alsahli

Abstract

The aetiology of epilepsy is multifactorial that disrupts the neuronal circuit through the generation of electric impulses and thus affects the cognitive, neurobiological and psychological systems. It is characterised by disturbances in motor and autonomic functions. Existing therapy focuses on curbing neuronal excitability but is symptomatic and ineffective in transposing the condition and restoring it to its normal state. Enzymes are an area of thrust that may facilitate the reversal process, chiefly the modulatory enzymes of the brain. The enzymatic regime is extensively exploited as a therapeutic option predominantly at a multidimensional level. This perspective elucidates some of the crucial enzymes researched that are beneficial for treating epilepsy along with the contributions to epileptic therapeutic regimens.

Keywords

Epilepsy · Enzymes · GABA-T · Carbonic anhydrase · Glutamine synthetase · mTOR · LDH

M. Mohammed (✉)
Department of Pharmaceutical Chemistry, KMCT College of Pharmacy, Mampara, Malappuram, Kerala, India

P. Prabitha
Department of Pharmaceutical Chemistry, JSS College of Pharmacy, JSS Academy of Higher Education & Research, Mysore, Karnataka, India

T. G. Alsahli
Department of Pharmacology, College of Pharmacy, Jouf University, Sakaka, Aljouf, Saudi Arabia

B. Mathew, D. G. T. Parambi (eds.), *Enzymes in Neurodegenerative Disorders*,
https://doi.org/10.1007/978-981-97-6822-6_5

5.1 Introduction

Epilepsy is a group of chronic non-communicable diseases with cognitive, neurobiological, psychological and social concerns, characterised by a long-term predisposition to produce epileptic seizures (Fisher et al. 2005). As a common neurological disorder affecting the worldwide population after Alzheimer's and Parkinson's diseases (MedStar Health 2019), epilepsy is broadly considered a disease of imbalanced neuronal network excitability due to modifications in synaptic or ionic transmission (Patel et al. 2019). Epilepsy can then be further characterised by different presenting features such as the age of onset, electroencephalogram (EEG) findings and neurologic examination. The symptoms of the disease significantly differ from patient to patient and are impacted mainly by the type of seizure or epilepsy. Major disturbances are found in motor function, autonomic function, sensory perception and behaviour. Sensory signs include paresthesias, atypical tastes or smells, flashing lights and auditory changes. Several autonomic symptoms like sweating, epigastric sensation and/or piloerection have also been reported. Some patients have experienced psychic symptoms including dysphasia, fear and hallucinations (Ha and Bellanger 2013). The therapeutic strategy of patients with epilepsy is predominantly focused on individualised symptoms with control in seizures in terms of frequency and severity. Epilepsy control using antiseizure medications or antiepileptic drugs (AEDs) is based on factors including efficacy, pharmacokinetics, serum levels, side effects, hormonal alteration and interactions between AEDs and other drugs (Cárdenas-Rodríguez et al. 2013). Non-pharmacological management of epilepsy includes surgical interventions, stimulation of the vagal nerve, use of a ketogenic diet and application of other alternative/complementary systems of medicine. Therapeutic alternatives in complementary systems include practices such as yoga, chiropractic, acupuncture, massage treatment, biofeedback through EEG, homeopathic treatment, aromatherapy and herbal remedies (Saxena and Nadkarni 2011). However, the mainstay approach for seizures is management by the use of AEDs that act via multiple biochemical mechanisms (Fig. 5.1). Some of the medications act via sodium channel and/or calcium channel blockade, opening neuronal potassium channels, augmenting gamma-aminobutyric acid (GABA) transmission through GABA-A receptors, blocking *N*-methyl-D-aspartic acid (NMDA) receptors through attachment to synaptic vesicle protein (SV2A) in the brain for modulating the release of synaptic neurotransmitter and inhibiting several underlying enzymes (Subbarao et al. 2023). The pharmacological intervention for epilepsy comprises more than 30 drugs on a timeline basis (Table 5.1).

The existence of several enzymes in the brain represents a sophisticated route for the activation and inactivation of numerous molecular pathways and subsequently interacts with the modulatory role(s) during the development of behavioural and pathological changes associated with epilepsy and neurological conditions. Such enzymes have rerouted as another course of action to the existing/developing therapeutic systems. The current chapter focuses on the major enzymes that are pivotal in the pathophysiology of various types of epilepsy.

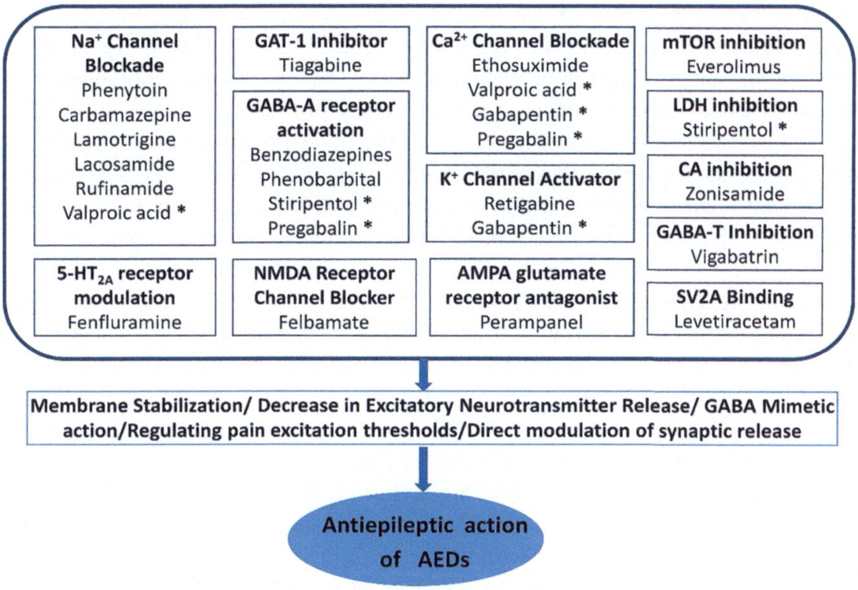

Fig. 5.1 Biochemical mechanisms of different antiepileptic drugs. AMPA, α-amino-3-hydroxy-5-methyl-4-isoxazolepropionic acid; CA, carbonic anhydrase; GABA-A, gamma-aminobutyric acid Type A; GABA-T, GABA transferase; GAT-1, GABA transporter type 1; LDH, lactate dehydrogenase; mTOR, mechanistic target of rapamycin; NMDA, *N*-methyl-D-aspartic acid; SV2A, synaptic vesicle protein; $5HT_{2A}$, 5-hydroxytryptamine (serotonin) Type 2A

Table 5.1 Drugs used for the treatment of epilepsy in a timeline

Years of introduction	List of drugs
1900–1950	Phenobarbital, phenytoin
1951–1985	Carbamazepine, diazepam, clobazam, clonazepam, midazolam, ethosuximide, primidone, valproic acid
1986–2000	Felbamate, fosphenytoin, lamotrigine, gabapentin, levetiracetam, lorazepam, oxcarbazepine, tiagabine, vigabatrin, topiramate, zonisamide
2001–2015	Eslicarbazepine acetate, perampanel, lacosamide, rufinamide, pregabalin, stiripentol
2015–2021	Brivaracetam, cenobamate, cannabidiol, everolimus, fenfluramine

5.2 Enzymes Responsible for Epilepsy

5.2.1 Gamma-Aminobutyric Acid Aminotransferase

GABA aminotransferase (GABA-AT), also known as GABA transferase (GABA-T), is chiefly responsible for the breakdown of GABA, the inhibitory neurotransmitter. When the levels of GABA go below the threshold, the increased neuronal excitation leads to convulsions. This inconsistency is rectified through GABA-T inhibition, which is taken as a new therapeutic alternative for epilepsy (Silverman 2018; Rimmer et al. 1988). GABA-T is a member of the subfamily II of α-type of pyridoxal 5′-phosphate (PLP)-dependent enzymes. The enzyme is a homodimer in which individual subunit comprises a catalytic site PLP that covalently is bound to Lys-329 using a Schiff's base (Storici et al. 2004). The nitrogen of Schiff's base lies almost in the same plane as that of the pyridine ring of PLP. It also interacts through a hydrogen bond with the pyridine oxygen (O_3), which implies the presence of a protonated aldimine (Storici et al. 2004). Moreover, the iron-sulphur clusters (2Fe-2S) assist in the cross-linking process of the subunits of the dimer.

Vigabatrin (gamma-vinyl GABA, GVG) is an AED that raises the GABA concentration in the brain through irreversible blockade of GABA-T (Sherif and Oreland 1995) (Fig. 5.2). The crystal structure of the drug, Vigabatrin with GABA-T depicts the manifestation of a ternary adduct between the drug, the co-factor PLP and the residue Lys-329 (Fig. 5.3a, b). Moreover, the electron density specifies the existence of four bonds connecting N to NZ. The density was found in proximity to Arg-192, perhaps demonstrating the carboxylate group of the inhibitor. The formed ternary adduct is well lodged in the enzyme active site with the NZ amino group. Thus, it forms a hydrogen bond with the main chain carbonyl of residue Ile-72 and the NH of the aldimine. This lies in the plane of the heterocyclic ring of pyridine forming another hydrogen bond with the O3 atom (Storici et al. 2004).

The process of GABA-T inactivation by Vigabatrin has been reported through different plausible mechanisms (Silverman 2018; Nanavati and Silverman 1991). In one mechanism (the most accepted), an initial azallylic isomerisation followed by elimination of γ-proton is reported, which is on par with the normal enzyme-substrate catalysis. Here, the inhibitor transforms into an electrophilic species that

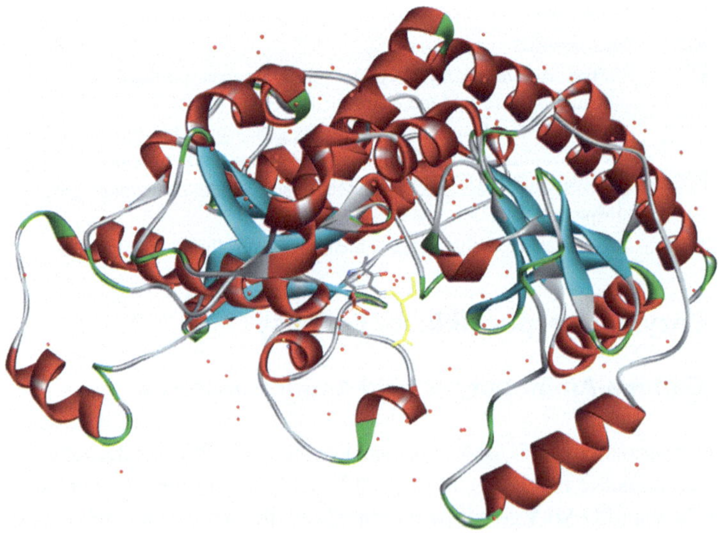

Fig. 5.2 Crystal structure of GABA-T with inhibitor, Vigabatrin (PDB ID:1OHW)

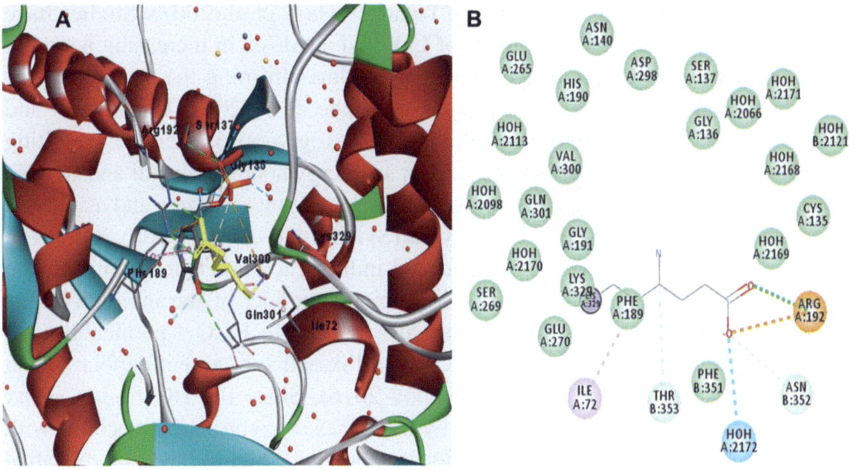

Fig. 5.3 (**a**) 3D interaction of Vigabatrin with GABA-aminotransferase enzymes and (**b**) 2D binding pose depiction of protein-ligand complex with co-factor PLP where Arg-192, Lys-329, Thr-353 and Ile-72 are the interacting amino acid residues

undergoes Michael addition at the carbon in the terminal of the vinyl group with Lys-329 (of the catalytic pocket) to yield an adduct. This adduct further undergoes tautomerism in several forms. The crystal structure of GABA-T with Vigabatrin demonstrates the geometry of one of these tautomers to be holding with a torsional angle of −169° around the N–CA aldimine linkage and an additional torsion angle of 1°. Both of these results in coplanarity of the pyridine ring with the aldimine linkage and a favourable hydrogen bond between the ionised NH^+ and $O3^-$ (Nanavati and Silverman 1991).

Another mechanism that involves the formation of enamine addition adducts has also been demonstrated. In this minor mechanism, the elimination of γ-proton occurs, but tautomerisation happens via the vinyl group. Subsequently, an enamine rearrangement follows to cause the binding of the inactivator to the PLP, present attached to the protein (Nanavati and Silverman 1991). This mechanism was reported with a study on an analogue of Vigabatrin, *cis*-3-aminocyclohex-4-ene-1-carboxylic acid (which is conformationally rigid) with GABA-T, which inhibited bond rotation and finally disrupted the Michael addition pathway (Choi et al. 2002).

5.2.2 Carbonic Anhydrase

Carbonic anhydrases (CAs) are known for their therapeutic potential for glaucoma, congestive heart failure, diuretics, intracranial hypertension and epilepsy (Aslam and Gupta 2023). These are a group of ubiquitous zinc-based enzymes that catalyse the hydration and dehydration of carbon dioxide-carbonic acid in a reversible mode. These enzymes are involved in the physiological and pathological courses implicated in maintaining the acid-base balance (Thiry et al. 2007). Studies have demonstrated the impact of low levels of CO_2 and alkalosis in increasing neuronal excitation and consequently triggering seizures, thus affirming the role of CAs in the pathological development of epilepsy (Ciccone et al. 2021; Ruusuvuori and Kaila 2014). During seizure attacks, the intensity of glycolysis increases resulting in the generation of metabolite lactic acid as an energy source (Enyedi and Czirják 2010; Yang et al. 2013). The translocation of the lactate into neurons and astrocytes is mediated through specific isoforms of CAs (Svichar and Chesler 2003). Moreover, CAs are reported to get involved in lactic acid-induced acidosis that can subside the triggered seizure (Yang et al. 2013). The excitatory effect of the inhibitory neurotransmitter GABA in convulsions may also be due to HCO_3^- permeability of GABA-A channels, which results in HCO_3^- efflux and subsequent depolarisation (Fujiwara-Tsukamoto et al. 2003; Khazipov 2016). Since the HCO_3^- ion efflux through GABA-A channels is mediated by CAs, these enzymes are associated with epileptogenesis (Staley et al. 1995). Various reports have highlighted that oxidative stress markedly contributes to the pathogenesis of convulsions. Overproduction of oxidative markers such as protein carbonylation, malondialdehyde, reactive oxygen species (ROS) and nitric oxide (NO) have progressively disturbed calcium homeostasis, which is crucial for neuronal existence (Svichar and Chesler 2003). A specific isoform, CA VII, plays in the cellular defence against oxidative damage via the cysteine residues (Deitmer et al. 2019).

CA inhibitors (CAI) are known to exhibit a good antiepileptic profile and the clinically successful molecules currently in use are acetazolamide, zonisamide, methazolamide and topiramate (Aggarwal et al. 2013). However, CAI acts as pan inhibitors, acting on several CA isoforms and the mechanism is not merely restricted to the CA inhibition. Furthermore, some of these have demonstrated unpleasant side effects. Thus, the research for newer agents to control such issues is currently under investigation.

5.2.3 Glutamine Synthetase (GS)

Glutamate is a key excitatory and potentially toxic neurotransmitter in the brain. Increased levels of extracellular glutamate and glutamate analogues in the brain can ultimately cause hyperexcitation, seizures and neuronal death. GS or glutamate ammonia ligase, a manganese-based octameric enzyme (Wedler et al. 1982; Denman and Wedler 1984), is known for the catalytic transformation of ammonia and glutamate to glutamine. The enzyme is highly localised in astrocytes and compartmentalised for numerous metabolic pathways in the brain that includes the glutamine-glutamate-GABA cycle and detoxification process of ammonia (Sandhu et al. 2021; Eid et al. 2012). Since GS is believed to be crucial for the metabolism of extracellular glutamate in astrocytes, enzyme deficiency is hypothesised as a plausible reason for the augmented extracellular levels of glutamate in the epileptogenic areas of the brain (Eid et al. 2004, 2008). The concept is further ascertained by isotopic tracer (C13) studies. The report depicts that the deceleration or disturbance of the glutamate-glutamine metabolic cycle in the epileptogenic hippocampus is accountable for the increased levels and reduced clearance of glutamate in Mesial Temporal Lobe Epilepsy (Petroff et al. 2002). Thus, it is clear that restoration of the enzyme through regulation/expression would prove beneficial in glutamate-glutamine conversion. Studies have revealed factors that decreased the expression of GS include genetic mutation in gene R324C, R341C (Häberle et al. 2005), hormones such as insulin (Khelil et al. 1990), nitrosative stress (Swamy et al. 2011), inflammatory mediators such as interleukin-1 β and tumour necrosis factor-α (Huang and O'Banion 1998) acidic pH (Nissim 1999), metal ion manganese (Deng et al. 2012) and exogenous agents like methionine sulphoxime (MSO) (Ronzio and Meister 1968), beta-*N*-oxalyl-L-alpha,beta-diaminopropionic acid (beta-L-ODAP) (Miller et al. 1993), etc. Similarly, factors increasing GS expression comprise hormones such as oetrogens (Blutstein et al. 2006), glucocorticoids (Patejunas and Young 1990), growth factor brain-derived neurotrophic factor (BDNF) (Dai et al. 2012), basic pH (Nissim 1999) and exogenous agents like resveratrol (dos Santos et al. 2006).

Neurotransmission of GABA also depends on GS inhibition as the precursor for GABA synthesis is glutamine. Inhibition of GS by MSO leads to declined production of neuronal GABA (Sonnewald et al. 1993) and impaired discharge of synaptic GABA under normal physiology (Liang et al. 2006). Moreover, MSO-mediated GS inhibition impacts GABAergic neurotransmission in a direct mode by modifying the composition of the GABA-A subunit (Cremer et al. 2010).

5.2.4 Lactate Dehydrogenase

Glycolysis ends with the generation of pyruvate, which is then transformed into lactate through lactate dehydrogenase (LDH). The formed lactate is shuttled through monocarboxylic acid transporters between astrocytes and neurons. Once entered into the neuron, LDH converts lactate back to pyruvate which goes over to the mitochondrion as feed into the Krebs cycle. The adenosine triphosphate (ATP) thus produced is released into the cytosol. Accumulated levels of ATP block ATP-sensitive potassium channels present on the neuronal plasma membrane which leads to rigorous neuronal discharge (implying epileptogenesis) (Rho 2015). However, low levels of ATP ultimately create an efflux of potassium ions to cause neuronal membrane hyperpolarisation and thus inhibit neuronal discharge. LDH inhibition thus inhibits the activation of ATP-sensitive potassium channels to bring down ATP production, rendering control in neuronal excitation (Sada et al. 2015).

Stiripentol, a clinically used AED, is known widely as a positive allosteric modulator of GABA-A receptor (Quilichini et al. 2006). It also acts as a barbiturate-like mechanism by binding to GABA-A receptor-dependent chloride channel (Frampton 2019). Moreover, it potentiates the transmission of GABA by enhancing the neurotransmitter release (Trojnar et al. 2005), decreasing its uptake into synapses and also reducing the breakdown of GABA by GABA transaminase (GABA-T) inhibition (European Medicines Agency 2010). In addition to these, stiripentol inhibits

LDH and thus helps to regulate the excitation of neurons (Sada et al. 2015). The drug is identified to demonstrate neuroprotective properties, thereby reducing injury produced by deprivation of oxygen-glucose and by excess levels of glutamate (Frampton 2019).

5.2.5 Mechanistic Target of Rapamycin (mTOR) Signalling Pathway

The mTOR is a serine-threonine protein kinase that forms part of two intracellular signalling complexes, namely, mTORC1 and mTORC2 (Ben-Sahra and Manning 2017; Saxton and Sabatini 2017). mTORC1 is the rapamycin-sensitive complex, whereas mTORC2 is comparatively the rapamycin-insensitive one. The influence of mTORC1, which comprises of modulating cellular growth, metabolism, proliferation and expression of ion-channel, is largely facilitated by ribosomal biogenesis and translation of protein (Griffith and Wong 2018). Genetic mutations in several modules of the mTOR signalling pathway have been identified to cause the overactivation of mTORC1. In turn, it may lead to various phenotypic changes in neuronal structure that include neuronal enlargement, alterations in morphology like size, proliferation, dendritic development and branching, the density of the dendritic spine, growth of axons, its regeneration and even disruption in cortical layering (Zhao et al. 2023). The disturbance also applies to synapses in terms of development and response. Such interference results in inevitable alterations in the neuronal network leading to the triggering of several diseases including epilepsy. Apart from mutations, neonatal hypoxia, traumatic injury to the brain, neuronal inflammation (Zhao et al. 2023), upregulation of tau protein (Griffith and Wong 2018; Mühlebner et al. 2019) and even status epilepticus also trigger the mTORC1 hyperactivation and thus initiate epileptogenic stimuli.

Rapamycin, a macrolide antibiotic and everolimus, a chemically modified variant of rapamycin, are established molecules as mTOR inhibitors. Rapamycin precisely obstructs mTOR activity but its effect on the two signalling complexes are different. The plausible mechanism of rapamycin in epilepsy is found to be via immunomodulation and anti-inflammatory responses, wherein T-cell migration is attenuated to relatively bring down the amplitude of neuronal excitation in seizures (Zhao et al. 2023). Studies have also demonstrated that everolimus treatment has effectively reduced the seizure frequency (Overwater et al. 2019).

Apart from these major ones, several minor enzymes have also been studied and implicated in the role(s) in convulsion. A few of them, but not exhaustive, are depicted in Table 5.2.

Table 5.2 List of minor enzymes with its involvement in epilepsy

Enzyme and role in epilepsy	References
Inosine triphosphatase (ITPA) • ITPA, involved in purine metabolism, prevents the incorporation of non-canonical purine nucleotides into nucleic acids • ITPA deficiency or genetic polymorphism in the corresponding gene leads to infantile epileptic encephalopathy	Zamzami (2022), Koga et al. (2020), Scala et al. (2022)
Calcineurin (CN) • CN, a calcium-dependent phosphatase, has a role in several neuronal processes such as learning and memory, modulating functions of receptors, neuronal excitation and inducing apoptosis • CN inhibitor and in vivo calcineurin A (CNA) knockdown experiments have depicted the activation of CN to induce epilepsy • Activation/upregulation of CN by NMDA receptor triggered calcium influx and downregulation of GABA receptor have also been reported for the trigger of seizures	Wang et al. (2009), Xiong et al. (2018), Ouyang et al. (2007), Kurz et al. (2003), Han et al. (2023)
Angiotensin-converting enzyme (ACE) • Angiotensin peptides are recognised to be implicated in the control of seizures. Specifically, ACE, angiotensin II (Ang II), Ang II type and type 2 (AT1 & AT2) receptors play crucial roles in the control of seizures • Thus, the treatment using ACE inhibitors (ACEi) can inhibit proconvulsant triggering exerted by Ang II to decrease the detrimental consequences resulting from epilepsy • Reports depict that ACEi like captopril, imidapril and quinapril were implicated in alleviating convulsions • A common agent as AED, carbamazepine, also acts to show ACE inhibition in patients with temporal lobe epilepsy	Tchekalarova and Georgiev (2005), Ivanova and Tchekalarova (2019), Krasniqi and Daci (2019) Dong et al. (2022), Pushpa et al. (2016), Almeida et al. (2012)

(continued)

Table 5.2 (continued)

Enzyme and role in epilepsy	References
Aconitase • Aconitases are iron-sulphur-based enzymes that mediate the reversible isomerisation of citrate and isocitrate through the intermediate, *cis*-aconitate in the TCA cycle • The enzyme is very vulnerable to oxidative species like free radicals, which renders the enzyme to be inactive. The generation of free radicals of oxygen and nitrogen species causes mitochondrial dysfunction which is frequently witnessed with seizures • Further recurrent convulsions lead to a cascade of events including downregulation of the aconitase enzyme, which follows out to decrease the expression of glutamate receptors. Thus, excess levels of glutamate would ultimately lead to neuronal excitability	Trujillo et al. (2010), Waldbaum and Patel (2010), Chan et al. (2019), Puttachary et al. (2015)
Adenosine kinase (ADK) • Adenosine is an inhibitory purine nucleoside that regulates brain activity. It is also known for its neuroprotective and anticonvulsant action. The levels of this modulator are mainly controlled by adenosine kinase (ADK), an enzyme that mediates the elimination of adenosine by phosphorylation to AMP • Increased levels of ADK in the epileptic brain directly trigger seizure generation by exhausting the quantity of adenosine. Such low levels of adenosine consequently decrease the DNA methylation process which also contributes to spontaneous focal seizures • ADK inhibitors (such as 5-iodotubercidin/5-ITU), indicated for chronic pain, inflammation and seizures in preclinical development, would in turn raise the levels of adenosine • Due to a lack of isoform selectivity and concerns like liver toxicity and cardiovascular effects, newer agents need to be substantialised	Boison (2006), During and Spencer (1992), Li et al. (2008), Williams-Karnesky et al. (2013), Jarvis (2019), Sandau et al. (2019), McGaraughty et al. (2005)

5.3 Conclusion

Epilepsy is considered to be a disease of imbalance in neuronal excitation which occurs due to reformed synaptic or ionic transmission. Synchronised neuronal electric discharge at an extreme level seeds out for epileptic seizures. Such explosive action potentials are clear manifestations of alteration in the neuronal signalling and networking pathways. Several methodologies, targets and research on neuronal pathways are currently addressed and established as measures for the therapy of epilepsy. As an add-on to the existing AEDs, reports implicate that investigations are underway for drugs with better therapeutic and safety profiles together with non-pharmacological interventions and alternative systems of medicine. In short, all the efforts employed intend to rectify the disrupted imbalance in neuronal excitability. Thus, the modulation in neuronal pathways would be restored, especially through enzymes. Signature enzymes such as GABA-T, carbonic anhydrase, mTOR signalling and several others have switched out a paradigm as a remedy for the disease and still many more are in exploration. Moreover, a resurgence of interest has been implicated in the involvement of metabolic pathways, transporters and linked enzymes such as LDH, aldehyde dehydrogenase and similar ones in the pathogenesis of epilepsy. These areas are still under research and need to be managed with the applicable scientific data and supporting technologies as multifaceted regimen for epilepsy management.

References

Aggarwal M, Kondeti B, McKenna R (2013) Anticonvulsant/antiepileptic carbonic anhydrase inhibitors: a patent review. Expert Opin Ther Pat 23:717–724

Almeida SS, Naffah-Mazzacoratti MG, Guimarães PB et al (2012) Carbamazepine inhibits angiotensin I-converting enzyme, linking it to the pathogenesis of temporal lobe epilepsy. Transl Psychiatry 2:e93

Aslam S, Gupta V (2023) Carbonic anhydrase inhibitors. StatPearls, Treasure Island

Ben-Sahra I, Manning BD (2017) mTORC1 signaling and the metabolic control of cell growth. Curr Opin Cell Biol 45:72–82

Blutstein T, Devidze N, Choleris E, Jasnow AM, Pfaff DW, Mong JA (2006) Oestradiol up-regulates glutamine synthetase mRNA and protein expression in the hypothalamus and hippocampus: implications for a role of hormonally responsive glia in amino acid neurotransmission. J Neuroendocrinol 18:692–702

Boison D (2006) Adenosine kinase, epilepsy and stroke: mechanisms and therapies. Trends Pharmacol Sci 27:652–658

Cárdenas-Rodríguez N, Coballase-Urrutia E, Rivera-Espinosa L, Romero-Toledo A, Sampieri AI, Ortega-Cuellar D, Montesinos-Correa H, Floriano-Sánchez E, Carmona-Aparicio L (2013) Modulation of antioxidant enzymatic activities by certain antiepileptic drugs (valproic acid, oxcarbazepine, and topiramate): evidence in humans and experimental models. Oxidative Med Cell Longev 2013:598493

Chan F et al (2019) The role of astrocytes in seizure generation: insights from a novel in vitro seizure model based on mitochondrial dysfunction. Brain 142(2):391–411. https://www.ncbi.nlm.nih.gov/pmc/articles/PMC6519661/. Accessed 4 Sep 2023

Choi S, Storici P, Schirmer T, Silverman RB (2002) Design of a conformationally restricted analogue of the antiepilepsy drug vigabatrin that directs its mechanism of inactivation of γ-aminobutyric acid aminotransferase. J Am Chem Soc 124:1620–1624

Ciccone L, Cerri C, Nencetti S, Orlandini E (2021) Carbonic anhydrase inhibitors and epilepsy: state of the art and future perspectives. Molecules 26:6380

Cremer CM, Bidmon H-J, Görg B, Palomero-Gallagher N, Escobar JL, Speckmann E-J, Zilles K (2010) Inhibition of glutamate/glutamine cycle in vivo results in decreased benzodiazepine binding and differentially regulated GABAergic subunit expression in the rat brain. Epilepsia 51:1446–1455

Dai M, Xia X-B, Xiong S-Q (2012) BDNF regulates GLAST and glutamine synthetase in mouse retinal Müller cells. J Cell Physiol 227:596–603

Deitmer JW, Theparambil SM, Ruminot I, Noor SI, Becker HM (2019) Energy dynamics in the brain: contributions of astrocytes to metabolism and pH homeostasis. Front Neurosci 13:1301

Deng Y, Xu Z, Xu B, Xu D, Tian Y, Feng W (2012) The protective effects of riluzole on manganese-induced disruption of glutamate transporters and glutamine synthetase in the cultured astrocytes. Biol Trace Elem Res 148:242–249

Denman RB, Wedler FC (1984) Association-dissociation of mammalian brain glutamine synthetase: effects of metal ions and other ligands. Arch Biochem Biophys 232:427–440
Dong X, Fan J, Lin D et al (2022) Captopril alleviates epilepsy and cognitive impairment by attenuation of C3-mediated inflammation and synaptic phagocytosis. J Neuroinflammation 19:226
dos Santos AQ, Nardin P, Funchal C, de Almeida LMV, Jacques-Silva MC, Wofchuk ST, Gonçalves C-A, Gottfried C (2006) Resveratrol increases glutamate uptake and glutamine synthetase activity in C6 glioma cells. Arch Biochem Biophys 453:161–167
During MJ, Spencer DD (1992) Adenosine: a potential mediator of seizure arrest and postictal refractoriness. Ann Neurol 32:618–624
Eid T, Thomas M, Spencer D, Rundén-Pran E, Lai J, Malthankar G, Kim J, Danbolt N, Ottersen O, De Lanerolle N (2004) Loss of glutamine synthetase in the human epileptogenic hippocampus: possible mechanism for raised extracellular glutamate in mesial temporal lobe epilepsy. Lancet 363:28–37
Eid T, Williamson A, Lee T-SW, Petroff OA, De Lanerolle NC (2008) Glutamate and astrocytes-key players in human mesial temporal lobe epilepsy?: glutamate and astrocytes. Epilepsia 49:42–52
Eid T, Behar K, Dhaher R, Bumanglag AV, Lee T-SW (2012) Roles of glutamine synthetase inhibition in epilepsy. Neurochem Res 37:2339. https://link.springer.com/article/10.1007/s11064-012-0766-5. Accessed 20 Aug 2023
Enyedi P, Czirják G (2010) Molecular background of leak K+ currents: two-pore domain potassium channels. Physiol Rev 90:559–605
European Medicines Agency (2010) https://www.ema.europa.eu/en/documents/product-information/diacomit-epar-productinformation_en.pdf. Accessed on 24 Aug 2023.
Fisher RS, van Emde BW, Blume W, Elger C, Genton P, Lee P, Engel J (2005) Epileptic seizures and epilepsy: definitions proposed by the International League Against Epilepsy (ILAE) and the International Bureau for Epilepsy (IBE). Epilepsia 46:470–472
Frampton JE (2019) Stiripentol: a review in Dravet syndrome. Drugs 79:1785–1796
Fujiwara-Tsukamoto Y, Isomura Y, Nambu A, Takada M (2003) Excitatory gaba input directly drives seizure-like rhythmic synchronization in mature hippocampal CA1 pyramidal cells. Neuroscience 119:265–275
Griffith JL, Wong M (2018) The mTOR pathway in treatment of epilepsy: a clinical update. Future Neurol 13:49–58
Ha H, Bellanger R (2013) Epilepsy: treatment and management. https://www.uspharmacist.com/article/epilepsy-treatment-and-management. Accessed 17 Aug 2023
Häberle J, Görg B, Rutsch F et al (2005) Congenital glutamine deficiency with glutamine synthetase mutations. N Engl J Med 353:1926–1933
Han Q-T, Yang W-Q, Zang C et al (2023) The toxic natural product Tutin causes epileptic seizures in mice by activating calcineurin. Signal Transduct Target Ther 8:101
Huang TL, O'Banion MK (1998) Interleukin-1 beta and tumor necrosis factor-alpha suppress dexamethasone induction of glutamine synthetase in primary mouse astrocytes. J Neurochem 71:1436–1442
Ivanova N, Tchekalarova J (2019) The potential therapeutic capacity of inhibiting the brain renin-angiotensin system in the treatment of co-morbid conditions in epilepsy. CNS Drugs 33:1101–1112
Jarvis MF (2019) Therapeutic potential of adenosine kinase inhibition—revisited. Pharmacol Res Perspect 7:e00506
Khazipov R (2016) GABAergic synchronization in epilepsy. Cold Spring Harb Perspect Med 6:a022764
Khelil M, Rolland B, Fages C, Tardy M (1990) Glutamine synthetase modulation in astrocyte cultures of different mouse brain areas. Glia 3:75–80
Koga Y, Tsuchimoto D, Hayashi Y, Abolhassani N, Yoneshima Y, Sakumi K, Nakanishi H, Toyokuni S, Nakabeppu Y (2020) Neural stem cell-specific ITPA deficiency causes neural depolarization and epilepsy. JCI Insight 5:e140229

Krasniqi S, Daci A (2019) Role of the angiotensin pathway and its target therapy in epilepsy management. Int J Mol Sci 20:726
Kurz JE, Rana A, Parsons JT, Churn SB (2003) Status epilepticus-induced changes in the subcellular distribution and activity of calcineurin in rat forebrain. Neurobiol Dis 14:483–493
Li T, Ren G, Lusardi T, Wilz A, Lan JQ, Iwasato T, Itohara S, Simon RP, Boison D (2008) Adenosine kinase is a target for the prediction and prevention of epileptogenesis in mice. J Clin Invest 118(2):571–582
Liang S-L, Carlson GC, Coulter DA (2006) Dynamic regulation of synaptic GABA release by the glutamate-glutamine cycle in hippocampal area CA1. J Neurosci 26:8537–8548
McGaraughty S, Cowart M, Jarvis MF, Berman RF (2005) Anticonvulsant and antinociceptive actions of novel adenosine kinase inhibitors. Curr Top Med Chem 5(1):43–58. https://doi.org/10.2174/1568026053386845
MedStar Health (2019) 5-common-neurological-disorders-and-how-to-identify-them. https://www.medstarhealth.org/blog/5-common-neurological-disorders-and-how-to-identify-them. Accessed 17 Aug 2023
Miller S, Nunn PB, Bridges RJ (1993) Induction of astrocyte glutamine synthetase activity by the lathyrus toxin beta-N-oxalyl-L-alpha,beta-diaminopropionic acid (beta-L-ODAP). Glia 7:329–336
Mühlebner A, Bongaarts A, Sarnat HB, Scholl T, Aronica E (2019) New insights into a spectrum of developmental malformations related to mTOR dysregulations: challenges and perspectives. J Anat 235:521–542
Nanavati SM, Silverman RB (1991) Mechanisms of inactivation of .gamma.-aminobutyric acid aminotransferase by the antiepilepsy drug .gamma.-vinyl GABA (vigabatrin). J Am Chem Soc 113:9341–9349
Nissim I (1999) Newer aspects of glutamine/glutamate metabolism: the role of acute pH changes. Am J Phys 277:F493–F497
Ouyang Y, Yang X-F, Hu XY, Erbayat-Altay E, Zeng L-H, Lee J-M, Wong M (2007) Hippocampal seizures cause depolymerization of filamentous actin in neurons independent of acute morphological changes. Brain Res 1143:238–246
Overwater IE, Rietman AB, van Eeghen AM, de Wit MCY (2019) Everolimus for the treatment of refractory seizures associated with tuberous sclerosis complex (TSC): current perspectives. Ther Clin Risk Manag 15:951–955
Patejunas G, Young AP (1990) Constitutive and glucocorticoid-mediated activation of glutamine synthetase gene expression in the developing chicken retina. J Biol Chem 265:15280–15285
Patel DC, Tewari BP, Chaunsali L, Sontheimer H (2019) Neuron-glia interactions in the pathophysiology of epilepsy. Nat Rev Neurosci 20:282–297
Petroff OAC, Errante LD, Rothman DL, Kim JH, Spencer DD (2002) Glutamate-glutamine cycling in the epileptic human hippocampus. Epilepsia 43:703–710
Pushpa VH, Poornima R, Kalabharathi HL, Satish AM (2016) Evaluation of anticonvulsant activity of ACE inhibitors (imidapril and quinapril) in experimentally induced animal models of epilepsy. Int J Basic Clin Pharmacol 5:798–801
Puttachary S, Sharma S, Stark S, Thippeswamy T (2015) Seizure-induced oxidative stress in temporal lobe epilepsy. Biomed Res Int 2015:1–20
Quilichini PP, Chiron C, Ben-Ari Y, Gozlan H (2006) Stiripentol, a putative antiepileptic drug, enhances the duration of opening of GABA-A receptor channels. Epilepsia 47:704–716
Rho JM (2015) Inhibition of lactate dehydrogenase to treat epilepsy. N Engl J Med 373:187–189
Rimmer E, Kongola G, Richens A (1988) Inhibition of the enzyme, GABA-aminotransferase in human platelets by vigabatrin, a potential antiepileptic drug. Br J Clin Pharmacol 25:251–259
Ronzio RA, Meister A (1968) Phosphorylation of methionine sulfoximine by glutamine synthetase. Proc Natl Acad Sci U S A 59(1):164–170. https://www.ncbi.nlm.nih.gov/pmc/articles/PMC286017/. Accessed 21 Aug 2023
Ruusuvuori E, Kaila K (2014) Carbonic anhydrases and brain pH in the control of neuronal excitability. Subcell Biochem 75:271–290

Sada N, Lee S, Katsu T, Otsuki T, Inoue T (2015) Epilepsy treatment. Targeting LDH enzymes with a stiripentol analog to treat epilepsy. Science 347:1362–1367

Sandau US, Yahya M, Bigej R, Friedman JL, Saleumvong B, Boison D (2019) Transient use of a systemic adenosine kinase inhibitor attenuates epilepsy development in mice. Epilepsia 60:615–625

Sandhu MRS, Gruenbaum BF, Gruenbaum SE, Dhaher R, Deshpande K, Funaro MC, Lee T-SW, Zaveri HP, Eid T (2021) Astroglial glutamine synthetase and the pathogenesis of mesial temporal lobe epilepsy. Front Neurol 12:665334

Saxena VS, Nadkarni VV (2011) Nonpharmacological treatment of epilepsy. Ann Indian Acad Neurol 14:148–152

Saxton RA, Sabatini DM (2017) mTOR signaling in growth, metabolism, and disease. Cell 168:960–976

Scala M, Wortmann SB, Kaya N et al (2022) Clinico-radiological features, molecular spectrum, and identification of prognostic factors in developmental and epileptic encephalopathy due to inosine triphosphate pyrophosphatase (ITPase) deficiency. Hum Mutat 43:403–419

Sherif F, Oreland L (1995) Effect of the GABA-transaminase inhibitor vigabatrin on exploratory behaviour in socially isolated rats. Behav Brain Res 72:135–140

Silverman RB (2018) Design and mechanism of GABA aminotransferase inactivators. Treatments for epilepsies and addictions. Chem Rev 118:4037–4070

Sonnewald U, Westergaard N, Schousboe A, Svendsen JS, Unsgård G, Petersen SB (1993) Direct demonstration by [13C]NMR spectroscopy that glutamine from astrocytes is a precursor for GABA synthesis in neurons. Neurochem Int 22:19–29

Staley KJ, Soldo BL, Proctor WR (1995) Ionic mechanisms of neuronal excitation by inhibitory GABAA receptors. Science 269:977–981

Storici P, De Biase D, Bossa F, Bruno S, Mozzarelli A, Peneff C, Silverman RB, Schirmer T (2004) Structures of γ-aminobutyric acid (GABA) aminotransferase, a pyridoxal 5′-phosphate, and [2Fe-2S] cluster-containing enzyme, complexed with γ-ethynyl-GABA and with the antiepilepsy drug vigabatrin. J Biol Chem 279:363–373

Subbarao BS, Silverman A, Eapen BC (2023) Seizure medications. StatPearls, Treasure Island

Svichar N, Chesler M (2003) Surface carbonic anhydrase activity on astrocytes and neurons facilitates lactate transport. Glia 41:415–419

Swamy M, Yusof WRW, Sirajudeen KNS, Mustapha Z, Govindasamy C (2011) Decreased glutamine synthetase, increased citrulline-nitric oxide cycle activities, and oxidative stress in different regions of brain in epilepsy rat model. J Physiol Biochem 67:105–113

Tchekalarova J, Georgiev V (2005) Angiotensin peptides modulatory system: how is it implicated in the control of seizure susceptibility? Life Sci 76:955–970

Thiry A, Dogné J-M, Supuran CT, Masereel B (2007) Carbonic anhydrase inhibitors as anticonvulsant agents. Curr Top Med Chem 7:855–864

Trojnar MK, Wojtal K, Trojnar MP, Czuczwar SJ (2005) Stiripentol. A novel antiepileptic drug. Pharmacol Rep 57:154–160

Trujillo M, Alvarez B, Souza JM, Romero N, Castro L, Thomson L, Radi R (2010) Chapter 3—Mechanisms and biological consequences of peroxynitrite-dependent protein oxidation and nitration. In: Ignarro LJ (ed) Nitric oxide, 2nd edn. Academic Press, San Diego, pp 61–102

Waldbaum S, Patel M (2010) Mitochondria, oxidative stress, and temporal lobe epilepsy. Epilepsy Res 88:23–45

Wang A, Chi Z, Wang S, Wang S, Sun Q (2009) Calcineurin-mediated GABAA receptor dephosphorylation in rats after kainic acid-induced status epilepticus. Seizure 18:519–523

Wedler FC, Denman RB, Gary Roby W (1982) Glutamine synthetase from ovine brain is a manganese(II) enzyme. Biochemistry 21(25):6389–6396. https://pubs.acs.org/doi/abs/10.1021/bi00268a011. Accessed 21 Aug 2023

Williams-Karnesky RL, Sandau US, Lusardi TA, Lytle NK, Farrell JM, Pritchard EM, Kaplan DL, Boison D (2013) Epigenetic changes induced by adenosine augmentation therapy prevent epileptogenesis. J Clin Invest 123:3552–3563

Xiong T-Q, Chen L-M, Tan B-H, Guo C-Y, Li Y-N, Zhang Y-F, Li S-L, Zhao H, Li Y-C (2018) The effects of calcineurin inhibitor FK506 on actin cytoskeleton, neuronal survival and glial reactions after pilocarpine-induced status epilepticus in mice. Epilepsy Res 140:138–147

Yang H, Wu J, Guo R, Peng Y, Zheng W, Liu D, Song Z (2013) Glycolysis in energy metabolism during seizures. Neural Regen Res 8:1316–1326

Zamzami MA (2022) Inosine triphosphate pyrophosphatase (ITPase): functions, mutations, polymorphisms and its impact on cancer therapies. Cells 11:384

Zhao W, Xie C, Zhang X, Liu J, Liu J, Xia Z (2023) Advances in the mTOR signaling pathway and its inhibitor rapamycin in epilepsy. Brain Behav 13:e2995

Role of Enzymes in Huntington's Disease

6

Malavila B. Nair, Gouri Shyna Rajesh, Della Grace Thomas Parambi, Arafa Musa, Tariq Ghazi Alsahli, Siju Ellickal Narayanan, Sandeep Bindra, and Bijo Mathew

Abstract

Huntington's disease (HD) is an autosomal disease affecting the striatum region of the brain. The gene responsible for HD is called Huntingtin. The onset of the disease depends on cysteine–adenine–guanine expansion. The striatum is the critical part of the brain affected by many movement disorders. Damage to the striatum can lead to uncontrolled movements of the face, trunk and limbs. These symptoms can worsen over time, then turn into abnormal and later into a life-threatening and fatal state. The enzymes responsible for these diseases are ubiquitin-modifying enzymes, phosphodiesterase and glutamic acid decarboxylase. There is no medicine that can treat, halt, slow, or reverse the progression of the disease. However, many interventions and medications are suggested by neurol-

M. B. Nair · G. S. Rajesh · S. Bindra
Department of Pharmaceutical Chemistry, Amrita School of Pharmacy, Amrita Vishwa Vidyapeetham, AIMS Health Sciences Campus, Kochi, India

D. G. T. Parambi
Department of Pharmaceutical Chemistry, Jouf University, Sakaka, Saudi Arabia

A. Musa
Department of Pharmacognosy, College of Pharmacy, Jouf University, Sakaka, Saudi Arabia

T. G. Alsahli
Department of Pharmacology, College of Pharmacy, Jouf University, Sakaka, Saudi Arabia

S. E. Narayanan
Department of Pharmacology, College of Pharmaceutical Sciences, Government Medical College, Kannur, India

B. Mathew (✉)
Department of Pharmaceutical Chemistry, Amrita School of Pharmacy, Amrita Vishwa Vidyapeetham, Kochi, Kerala, India

B. Mathew, D. G. T. Parambi (eds.), *Enzymes in Neurodegenerative Disorders*,
https://doi.org/10.1007/978-981-97-6822-6_6

ogists, psychiatrists and experts in HD to ease the anxiety, depression and troublesome comorbidities of HD.

Keywords

Huntington's disease · Neurodegenerative disease · Polyglutamine · Pathogenesis · FDA-approved drugs

6.1 Introduction

Huntington's disease (HD), also known as Huntington's chorea, is a rare autosomal dominant neurodegenerative disease caused by the repetition of glutamine trinucleotide cysteine–adenine–guanine (CAG) in the huntingtin gene (HTTg). DNA error that is responsible for HD is found in a gene huntingtin. The normal HTTg has CAG repeats ranging from 10 to 35, whereas the mutated HTTg contains CAG repeats of more than 36 on the shorter arm of chromosome 4p16.3 and results in polyglutamine (polyQ) around the N terminus of HTTg (Tobore 2019).

HD has psychiatric and motor symptoms which include loss of coordination, chorea depression, psychosis and obsessive-compulsive disorder. The onset of symptoms of Huntington's disease (HD) depends on the CAG expansion (Jimenez-Sanchez et al. 2017), that is, the adult onset of HD is associated with 40–49 CAG repeats with symptoms (motor symptoms) appearing between 30 and 50 years and juvenile onset if it is more than 50 CAG repeats where the symptoms start before the 2nd decade of life (Ferrer 2016).

6.2 Pathophysiology of HD

HD shares a common genetic background, but the degradation of brain cells in the cerebral cortex and basal ganglia is directly associated with motor and psychiatric symptoms in HD patients (Cepeda and Tong 2018). The striatum (caudate and putamen) present within the basal ganglia consists of the dorsal and ventral striatum in which the dorsal striatum is responsible for motor and cognitive functions (Young 2023).

Misfolded proteins can be broken down by the ubiquitin-proteasome system (UPS). Due to the mutant huntingtin's resistance to the UPS, aberrant inclusion bodies accumulate in neurons and glial cells (Fletcher and Mabbitt 2023). To minimise the buildup of mHTT, many neurons and glial cells can raise their chaperone protein load and protein degradation machinery. However, medium spiny neurons (MSNs) are unable to do this, which causes the striatum to atrophy and neurotoxicity to develop.

The red area in the brain is the "striatum" which play a major role in the mood, behaviour and movement (Fig. 6.1).

MSNs predominate in the striatum which receives modulatory dopaminergic and excitatory glutaminergic signals and releases the inhibitory neurotransmitter

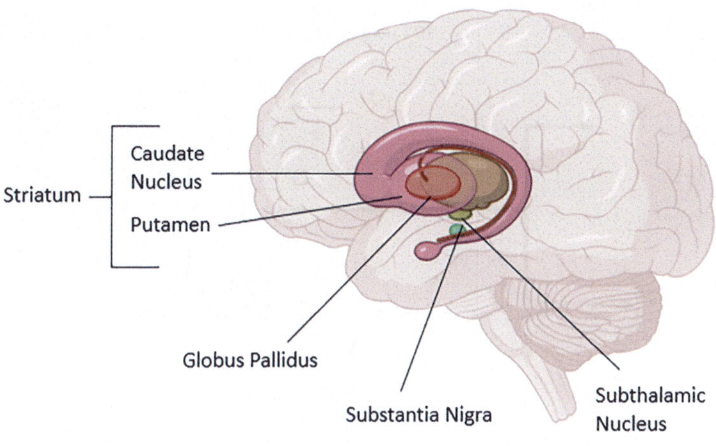

Fig. 6.1 Pathophysiology of the HD brain

gamma-aminobutyric acid (GABA) in response to these signals (Matz and Spocter 2022).

Type 1 dopaminergic receptors are expressed via the direct pathway of the corticostriatal system, while type 2 dopaminergic receptors are expressed through the indirect pathway (Calabresi et al. 2014).

Direct pathway: GABA neurons connected to D1 receptors in the striatum transmit inhibitory signals to the globus pallidus and substantia nigra reticular, which activates the thalamus through glutamatergic neurons, resulting in enhanced motor and behavioural function (Leisman and Melillo 2013).

An indirect pathway: GABA neurons in the striatum transmit inhibitory signals to the globus pallidum, which then proceed to the subthalamic nucleus. The globus pallidum and thalamus are inhibitory stimulated by the subthalamic nucleus's activity through glutaminergic neurons, which causes hypokinetic function (Leisman and Melillo 2013).MSNs are the cells with D2 receptors that are more selectively vulnerable to the effects of mHTT (Matz and Spocter 2022; McColgan and Tabrizi 2018). The frontal cortex sends information to the basal ganglia through five loops: the frontal eye field for eye movements, the anterior cingulate for motivation, the dorsiventral prefrontal cortex for coordination and execution and the premotor cortex for movement. These loops collectively are in charge of an individual's normal behavioural function (Leisman and Melillo 2013). While the indirect pathway is linked to the regulation of movement and cognitive processes, the direct pathway is linked to enhanced motor and behavioural function (Leisman and Melillo 2013; Hsu et al. 2018). The imbalance between the inhibitory GABAergic and the excitatory glutamatergic systems in the cortex is responsible for motor and behavioural dysfunction in HD patients (Hsu et al. 2018).

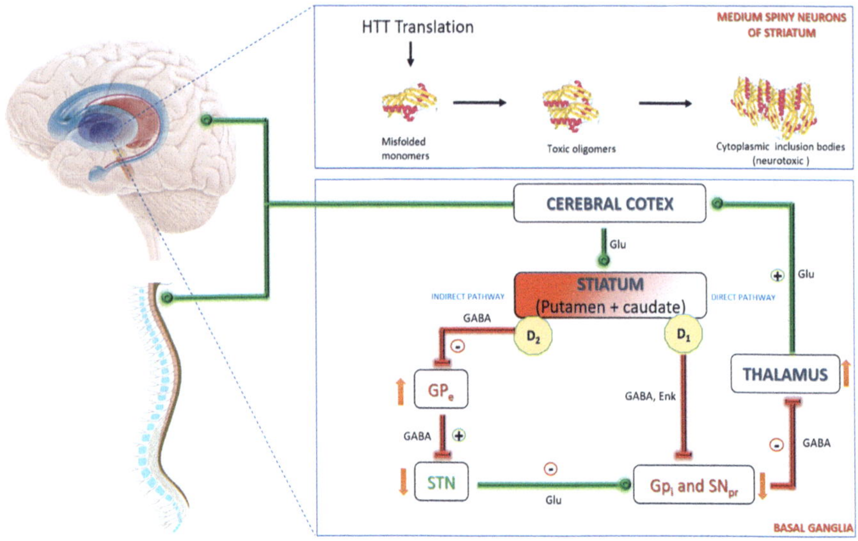

Protein aggregation leads to an imbalance in direct and indirect pathways. Neurotransmitters: *Glu* **glutamate,** *GABA* **gamma amino butyric acid,** *Enk* **enkephalin. Nuclei:** *Gpe* **globus pallidus externa,** Gp_i **globus pallidus interna,** *STN* **subthalamic nucleus,** SN_{pr} **substantia nigra pars reticulata**

The loss of striatal cells, particularly D2 dopaminergic neurons, is caused by the accumulation of cytoplasmic inclusion bodies in MSNs. This inhibition of the substantia nigra and globus pallidus interna follows. The thalamus receives inhibitory signals from these regions, and this further stimulates the motor and cognitive parts of the cortex, leading to the symptoms associated with HD.

6.3 HD Effect on the Central Nervous System (CNS)

HD can cause the face, torso and limbs to move quickly and uncontrollably in its early stages. People may initially combine these anomalous involuntary movements with intentional ones, causing the abnormal movements hardly perceptible. However, over the years, the movements become more noticeable.

Muscles can twitch quickly and sharply, occasionally causing the arms or another body part to jerk repeatedly.

People may have a puppet-like gait or an oversized, jaunty gait. They could squint, wave their limbs and blink more frequently. Slow, disorganised movements develop. Eventually, the entire body is impacted, making it very difficult to eat, speak, swallow, walk or even sit still (https://rarediseases.org/rare-diseases/huntingtons-disease).

Before or during the first episode of the aberrant motions, mental alterations frequently take place. Initially, these changes are minimal. Individuals may progressively become nervous, impatient and excitable. They could get disinterested in their routines. They might become impetuous, lose their cool, experience depressive episodes or start acting promiscuous. People with advanced illness have severe dementia and are bedridden. Care in a nursing home or full-time help is required. Typically, symptoms start to manifest 13–15 years later (https://rarediseases.org/rare-diseases/huntingtons-disease).

6.4 Enzymes Related to HD

1. *Ubiquitin-Modifying Enzymes*: The 76 amino acid complex associated with the ubiquitin-proteasome system (UPS) plays a role in a non-lysosomal proteolytic pathway and is required for the release of neurotransmitters, the breakdown of misfolded intracellular protein and the correct functioning of neurons (Momtaz et al. 2020). E1 ubiquitin-activating enzymes, E2 ubiquitin-conjugating enzymes and E3 ubiquitin ligase enzymes attach to the target protein to use it as a target for disintegration. Ubiquitin is a protein. Because wHTT is ubiquitinated at other sites but mHTT is ubiquitinated at K6 and K9, the dysfunction of the ubiquitin system results in the accumulation of mHTT as abnormal inclusion bodies in the cytoplasm of neurons and glial cells (Sap et al. 2023). These inclusion bodies are not effectively degraded. Soluble mHTT is better degraded by ubiquitin ligase enzymes such as HOIP, UBR5 and E3 ubiquitin ligase UBE3A. The ligase CHIP, through chaperone-mediated refolding, improves the solubilisation and degradation of mHTT (Sap et al. 2019).
2. *Phosphodiesterase (PDEs)*: These are the enzymes responsible for catabolism and cAMP and cGMP homeostasis maintenance (Hakim-Eshed et al. 2020). The catabolic enzyme phosphodiesterase is responsible for hydrolysing cAMP and cGMP, two cyclic nucleotides (Delhaye and Bardoni 2021). Cyclic nucleotides are the secondary messengers (Hakim-Eshed et al. 2020) associated with many cellular processes such as memory and neurotransmission through various signalling pathways like protein kinase A and extracellular regulated kinase and target cAMP response element binding protein (CREB). Neurotransmitter modulator brain-derived neurotrophic factor (BDNF) is depleted when phosphodiesterase inactivates cyclic nucleotide, which is necessary for neuronal survival (Cardinale and Fusco 2018). A CREB target gene called huntingtin mutation is connected to dysregulation in transcription, postsynaptic signalling and a cellular transport mechanism in HD patients, as seen by the decrease in BDNF (Cardinale and Fusco 2018).
3. *Glutamic acid decarboxylase (GAD)*: GABA is the primary inhibitory neurotransmitter essential for the brain's optimum equilibrium of synaptic excitement and inhibition (Bathina and Das 2015). The neurotransmitter enzyme responsible for the generation of GABA is known as GAD, and it exists in the inhibitory neurons of the dorsal striatum and cortex (Hsu et al. 2018). Altered levels in this neurotransmitter enzyme are associated with hyperkinetic conditions and the development of HD symptoms in patients (Hsu et al. 2018; Lee et al. 2019). The activity of GAD was found to be reduced by 65% in the puta-

men and a slight reduction in the frontal cortex in the experimental study done by J.-Y. Wu, E.D Bird, Department of Cell Biology Addenbrooke's Hospital Cambridge, England, in 1979. This reduction in GAD activities is due to the loss of neurons containing GABA which results in a decrease in GAD molecules (Bird and Iversen 1974).

6.5 FDA-Approved Drugs

Although there currently is no cure for HD, some treatments aim at alleviating its symptoms.

Various HD symptoms are treated with medications such as tetrabenazine, antipsychotics, dopamine receptor blocking or depleting medicines, neuroleptic pharmaceuticals (taken off-label) and antidepressants. For many years, tetrabenazine was widely used to treat hyperkinetic movement disorders, particularly dystonias and tics, as well as chorea in many nations. Neurons are where neurotransmitters are made. Afterward, these neurotransmitters are retained in vesicles and released at a later time. As a result, there are more neurotransmitters within the cell and fewer in the synaptic cleft, allowing for a controlled, progressive release of the neurotransmitters across the plasma membrane. The mechanism also shields the neuron from the potentially harmful consequences of dangerously elevated neurotransmitter levels, as well as from leakage or destruction by intraneuronal metabolism. The primary locations of the vesicular monoamine transporter-2 (VMAT-2) include the stomach, blood cells, adrenal medulla histaminergic cells and the CNS. Serotonin, dopamine, histamine and norepinephrine are all transported by the VMAT-2 into vesicles for storage (Yero and Rey 2008).

An antipsychotic drug called haldol (haloperidol) is used to treat severe behavioural issues such hyperactivity and violent and explosive behaviour, as well as psychotic disorders. By reducing the effects of dopamine and increasing its turnover, haldol can reduce the chorea symptoms associated with HD (https://huntingtonsdiseasenews.com/approved-treatments-for-huntingtons-disease-2).

The underlying mechanism of chorea was thought to be the activation of DA receptors, which in turn produced dyskinesias. Supporting evidence was found in HD patient investigations, which showed that a decrease in DA causes akinesia while an increase in DA release causes chorea (Chen et al. 2013).

The use of antipsychotics and tetrabenazine together is controversial since the antipsychotic medication negates the effects of tetrabenazine.

6.6 Comorbidities of HD

The study found that individuals with adult-onset HD had a higher incidence rate of many comorbidities, including dementia, obsessive-compulsive disorder, acute psychotic episodes, communication problems, depression, anxiety, pneumonia, constipation and fractures, as compared to controls. In patients with juvenile-onset HD, constipation, acute respiratory symptoms and seizures were more prevalent (https://rarediseases.org/rare-diseases/huntingtons-disease).

6.7 Conclusion

HD is a neurological disorder that completely alters the brain chemistry. This disorder disturbs the person as well as the people surrounding them completely. At 20 years after the symptom onset, surface death is bound to happen. HD has no known cure, and significant efforts to slow or halt its progression have not been developed. More light has been shed on this disease in recent years. Research has been going on in order to understand the exact reason and as to why this happens. In the coming years, we may experience some kind of cure for the same.

References

Bathina S, Das UN (2015) Brain-derived neurotrophic factor and its clinical implications. Arch Med Sci 6:1164–1178. https://doi.org/10.5114/aoms.2015.56342

Bird ED, Iversen LL (1974) Huntington's chorea: post-mortem measurement of glutamic acid decarboxylase, choline acetyltransferase and dopamine in Basal ganglia. Brain 97(3):457. https://doi.org/10.1093/brain/97.3.457

Calabresi P, Picconi B, Tozzi A, Ghiglieri V, Di Filippo M (2014) Direct and indirect pathways of basal ganglia: a critical reappraisal. Nat Neurosci 17(8):1022–1030. https://doi.org/10.1038/nn.3743

Cardinale A, Fusco FR (2018) Inhibition of phosphodiesterases as a strategy to achieve neuroprotection in Huntington's disease. CNS Neurosci Ther 24(4):319–328. https://doi.org/10.1111/cns.12834

Cepeda C, Tong X (2018) Huntington's disease: from basic science to therapeutics. CNS Neurosci Ther 24(4):247–249. https://doi.org/10.1111/cns.12841

Chen JY, Wang EA, Cepeda C, Levine MS (2013) Dopamine imbalance in Huntington's disease: a mechanism for the lack of behavioral flexibility. Front Neurosci 7:114. https://doi.org/10.3389/fnins.2013.00114

Delhaye S, Bardoni B (2021) Role of phosphodiesterases in the pathophysiology of neurodevelopmental disorders. Mol Psychiatry 26(9):4570–4582. https://doi.org/10.1038/s41380-020-00997-9

Ferrer I (2016) Mini-symposium: Huntington's disease - introduction. Brain Pathol 26(6):724–725. https://doi.org/10.1111/bpa.12431

Fletcher AJ, Mabbitt PD (2023) Editorial: Reviews in ubiquitin signaling: 2022. Front Mol Biosci 10:1275393. https://doi.org/10.3389/fmolb.2023.1275393

Hakim-Eshed V, Boulos A, Cohen-Rosenzweig C, Yu-Taeger L, Ziv T, Kwon YT, Riess O, Phuc Nguyen HH, Ziv NE, Ciechanover A (2020) Site-specific ubiquitination of pathogenic huntingtin attenuates its deleterious effects. Proc Natl Acad Sci 117(31):18661–18669. https://doi.org/10.1073/pnas.2007667117

Hsu Y-T, Chang Y-G, Chern Y (2018) Insights into $GABA_A$ ergic system alteration in Huntington's disease. Open Biol 8(12):180165. https://doi.org/10.1098/rsob.180165

Jimenez-Sanchez M, Licitra F, Underwood BR, Rubinsztein DC (2017) Huntington's disease: mechanisms of pathogenesis and therapeutic strategies. Cold Spring Harb Perspect Med 7(7):a024240. https://doi.org/10.1101/cshperspect.a024240

Lee S-E, Lee Y, Lee GH (2019) The regulation of glutamic acid decarboxylases in GABA neurotransmission in the brain. Arch Pharm Res 42(12):1031–1039. https://doi.org/10.1007/s12272-019-01196-z

Leisman G, Melillo R (2013) Clinical motor and cognitive neurobehavioral relationships in the basal ganglia. In: Basal ganglia - an integrative view. IntechOpen, London. https://doi.org/10.5772/55227

Matz OC, Spocter M (2022) The effect of Huntington's disease on the basal nuclei: a review. Cureus. https://doi.org/10.7759/cureus.24473

McColgan P, Tabrizi SJ (2018) Huntington's disease: a clinical review. Eur J Neurol 25(1):24–34. https://doi.org/10.1111/ene.13413

Momtaz S, Memariani Z, El-Senduny FF, Sanadgol N, Golab F, Katebi M, Abdolghaffari AH, Farzaei MH, Abdollahi M (2020) Targeting ubiquitin-proteasome pathway by natural products: novel therapeutic strategy for treatment of neurodegenerative diseases. Front Physiol 11:361. https://doi.org/10.3389/fphys.2020.00361

Sap KA, Guler AT, Bezstarosti K, Bury AE, Juenemann K, Demmers, Jeroen AA, Reits EA (2019) Global proteome and ubiquitinome changes in the soluble and insoluble fractions of Q175 Huntington mice brains. Mol Cell Proteomics 18(9):1705–1720. https://doi.org/10.1074/mcp.RA119.001486

Sap KA, Geijtenbeek KW, Schipper-Krom S, Guler AT, Reits EA (2023) Ubiquitin-modifying enzymes in Huntington's disease. Front Mol Biosci 10:1107323. https://doi.org/10.3389/fmolb.2023.1107323

Tobore TO (2019) Towards a comprehensive understanding of the contributions of mitochondrial dysfunction and oxidative stress in the pathogenesis and pathophysiology of Huntington's disease. J Neurosci Res 97(11):1455–1468. https://doi.org/10.1002/jnr.24492

Yero T, Rey JA (2008) Tetrabenazine (xenazine), an FDA-approved treatment option for Huntington's disease–related chorea. P T 33(12):690–694

Young CB (2023) Neuroanatomy, basal ganglia. In: StatPearls. StatPearls Publishing, Treasure Island

Role of Enzymes in Parkinson's Disease 7

Aathira Sujathan Nair, Sonia Dangi, Arun K. Thomas,
and Mahendran Sekar

Abstract

Parkinson's disease (PD) is a neurodegenerative condition marked by a steady loss of dopaminergic neurons in the brain's substantia nigra. Although the precise origin of PD is still unknown, new research indicates that enzymes are important in the pathophysiology of this crippling illness. We will examine several enzymes linked to PD and how they affect the course of the illness in this chapter. We will discuss the misregulation of enzymes related to mitochondrial dysfunction, oxidative stress, protein aggregation, and dopamine metabolism. Knowing how enzymes function in PD might help create new therapy approaches that target these enzymatic pathways.

Keywords

Parkinson's disease · Neuroinflammation · Dopamine · Protein aggregation

A. S. Nair (✉) · M. Sekar (✉)
School of Pharmacy, Monash University Malaysia, Subang Jaya, Selangor, Malaysia

S. Dangi
Jeffrey Cheah School of Medicine and Health Sciences, Monash University Malaysia, Subang Jaya, Selangor, Malaysia

A. K. Thomas
Department of Pharmacology, Annoor Dental College, Muvattupuzha, Kerala, India

B. Mathew, D. G. T. Parambi (eds.), *Enzymes in Neurodegenerative Disorders*,
https://doi.org/10.1007/978-981-97-6822-6_7

7.1 Introduction

Parkinson's disease (PD) is a complex neurodegenerative disorder that significantly impacts the lives of those affected, as well as their families and caregivers (Weiner et al. 2013). Named after the English physician James Parkinson, who first described the condition in 1817, PD is characterized by a range of motor and non-motor symptoms that progressively worsen over time. While the exact cause of PD remains elusive, advances in neuroscience have shed light on its pathological features, providing a foundation for understanding and treating the condition (Olanow et al. 2011).

7.1.1 Overview of PD

PD primarily affects the central nervous system, particularly the substantia nigra, a region in the brain responsible for producing dopamine—a neurotransmitter crucial for coordinating smooth and controlled movements. The hallmark motor symptoms of PD emerge when a significant portion of dopamine-producing neurons in the substantia nigra degenerate. These symptoms typically include tremors, bradykinesia (slowness of movement), rigidity, and postural instability (Rao et al. 2006).

7.1.1.1 Tremors

One of the most recognizable signs of PD is the presence of resting tremors, which usually start in the hands and can later affect other parts of the body. These tremors are often described as a rhythmic back-and-forth movement and are most pronounced when the individual is at rest (Helmich et al. 2012).

7.1.1.2 Bradykinesia

Bradykinesia refers to slowness of movement and a general reduction in spontaneous activities. Individuals with PD may find it challenging to initiate and complete everyday movements, such as walking, turning, or getting out of a chair. This symptom significantly impacts the quality of life for those with PD (Berardelli et al. 2001).

7.1.1.3 Rigidity

Rigidity involves stiffness of the muscles, making it difficult for individuals with PD to flex and extend their limbs smoothly. This stiffness contributes to the characteristic shuffling gait observed in many PD patients (Prochazka et al. 1997).

7.1.1.4 Postural Instability

As the disease progresses, individuals with PD may experience difficulties maintaining their balance and posture. This can lead to an increased risk of falls, which is a major concern for both patients and caregivers.

In addition to these primary motor symptoms, PD also manifests as various non-motor symptoms, further complicating the clinical picture. These non-motor symptoms may include cognitive impairment, mood disorders (such as depression and anxiety), sleep disturbances, and autonomic dysfunction (Palakurthi and Burugupally 2019).

7.1.1.5 Cognitive Impairment

While not all individuals with PD develop cognitive impairment, a significant portion may experience difficulties with memory, attention, and executive functions as the disease progresses. In some cases, PD can lead to dementia, further impacting the individual's ability to perform daily activities (Marsh 2013).

7.1.1.6 Mood Disorders

Depression and anxiety are common non-motor symptoms associated with PD. The emotional impact of living with a chronic and progressive condition, coupled with changes in brain chemistry, can contribute to these mood disorders (Menza et al. 2010).

7.1.1.7 Sleep Disturbances

Many individuals with PD experience disruptions in sleep patterns, including insomnia and disturbances in rapid eye movement sleep. These sleep disturbances can exacerbate other symptoms and negatively affect overall well-being (Palma and Kaufmann 2018).

7.1.1.8 Autonomic Dysfunction

PD can affect the autonomic nervous system, leading to symptoms such as orthostatic hypotension (a sudden drop in blood pressure upon standing), constipation, and urinary problems.

The onset of PD is typically gradual, and the progression of symptoms varies among individuals. Early diagnosis is challenging due to the absence of definitive biomarkers, and diagnosis is often based on clinical evaluation and the presence of specific motor symptoms (Kouli et al. 2018).

7.1.2 Pathological Features of PD

Understanding the pathological features of PD is crucial for developing targeted therapies and interventions. The primary neuropathological hallmark of PD is the presence of Lewy bodies—abnormal protein aggregates—within certain brain cells. These Lewy bodies are mainly composed of alpha-synuclein, a protein that normally plays a role in neurotransmitter release (Gómez-Benito et al. 2020).

7.1.2.1 Alpha-Synuclein Accumulation

In PD, alpha-synuclein undergoes abnormal aggregation, forming insoluble clumps within neurons. The exact mechanisms leading to alpha-synuclein aggregation are not fully understood, but genetic and environmental factors may contribute to this process (DeMaagd and Philip 2015).

7.1.2.2 Neuronal Degeneration

The presence of Lewy bodies is associated with the degeneration of dopamine-producing neurons in the substantia nigra. As these neurons die off, the levels of dopamine in the brain decrease, leading to the motor symptoms characteristic of PD (Wang et al. 2015).

7.1.2.3 Neuroinflammation

In addition to alpha-synuclein pathology, there is evidence of neuroinflammation in the brains of individuals with PD. Microglia, the immune cells of the central nervous system, become activated and contribute to the inflammatory response. This inflammatory process may exacerbate neuronal damage and contribute to the progression of the disease (Alexander 2004).

7.1.2.4 Spread of Pathology

PD is not limited to the substantia nigra; as the disease progresses, Lewy bodies and neurodegeneration can spread to other regions of the brain. This widespread involvement may explain the diverse range of motor and non-motor symptoms observed in individuals with PD.

Research into the underlying causes of PD is ongoing, and a combination of genetic, environmental, and lifestyle factors is believed to contribute to an individual's susceptibility to the disease. Some genetic mutations have been identified as risk factors for PD, although the majority of cases are considered sporadic, with no clear genetic predisposition (Zahoor et al. 2018).

7.1.3 Current Treatment Approaches and Limitations

While there is currently no cure for PD, several treatment approaches aim to alleviate symptoms and improve the quality of life for affected individuals. The primary focus of treatment is to restore or replace dopamine in the brain, as the motor symptoms of PD are directly linked to dopamine deficiency.

7.1.3.1 Dopamine Replacement Therapy

The most common and effective treatment for PD is dopamine replacement therapy. Levodopa, a precursor to dopamine, is converted to dopamine in the brain and helps alleviate motor symptoms. However, the long-term use of levodopa is associated with the development of motor fluctuations and dyskinesias—abnormal, involuntary movements (Choi and Horner 2019).

7.1.3.2 Dopamine Agonists

Dopamine agonists are drugs that mimic the action of dopamine in the brain. These medications can be used alone or in combination with levodopa. While they provide relief from motor symptoms, they may also be associated with side effects such as hallucinations and impulse control disorders (Schapira 2011).

7.1.3.3 MAO-B Inhibitors and Catechol-O-Methyltransferase (COMT) Inhibitors

Monoamine oxidase type B (MAO-B) inhibitors and COMT inhibitors are other classes of drugs used to prolong the effects of levodopa by preventing its breakdown. These medications are often used as adjuncts to levodopa therapy (Benabid 2003).

7.1.3.4 Deep Brain Stimulation (DBS)

In cases where medications alone do not provide sufficient symptom control, DBS may be considered (Niedzielska et al. 2016).

7.2 Oxidative Stress

Oxidative stress, characterized by an imbalance between reactive oxygen species (ROS) production and antioxidant defenses, is implicated in the pathogenesis of various neurodegenerative disorders, including PD. This comprehensive exploration delves into the intricate relationship between enzymes and oxidative stress, with a specific focus on PD (Lin and Beal 2006).

7.2.1 Oxidative Stress and PD

PD, a progressive neurodegenerative disorder, is marked by the selective loss of dopaminergic neurons in the substantia nigra. Oxidative stress emerges as a key player in the multifaceted etiology of PD. ROS, including superoxide radicals and hydrogen peroxide, inflict damage on cellular components, contributing to neuronal dysfunction and death (Checa and Aran 2020).

7.2.2 Enzymatic Sources of ROS

7.2.2.1 NADPH Oxidase

NADPH oxidase, a membrane-bound enzyme complex, is a prominent source of ROS in various cell types, including neurons. In PD, dysregulation of NADPH oxidase activity leads to an excessive generation of superoxide radicals, amplifying oxidative stress and exacerbating neurodegeneration (Pratomo et al. 2021).

7.2.2.2 Xanthine Oxidase

Xanthine oxidase, an enzyme involved in purine metabolism, also contributes to ROS production. Under pathological conditions, such as inflammation or ischemia, xanthine oxidase activity is heightened, leading to an increased generation of superoxide radicals. This mechanism further fuels the oxidative milieu in PD (Younus 2018).

7.2.3 Antioxidant Enzymes

7.2.3.1 Superoxide Dismutase (SOD)

SOD is a crucial antioxidant enzyme responsible for catalyzing the dismutation of superoxide radicals into hydrogen peroxide and oxygen. In PD, SOD activity is compromised, diminishing the cell's ability to neutralize superoxide radicals and contributing to the oxidative burden (Ursini and Bindoli 1987).

7.2.3.2 Glutathione Peroxidase

Glutathione peroxidase, a selenium-dependent enzyme, plays a pivotal role in detoxifying hydrogen peroxide and lipid hydroperoxides. In PD, reduced glutathione peroxidase activity compromises the cell's defense against peroxidative damage, amplifying oxidative stress and neuronal vulnerability (Nandi et al. 2019)

7.2.3.3 Catalase

Catalase is an enzyme localized in peroxisomes, responsible for decomposing hydrogen peroxide into water and oxygen. In PD, a decline in catalase activity impairs the cell's ability to eliminate hydrogen peroxide, contributing to the accumulation of this potent oxidant and exacerbating oxidative stress (Rahal et al. 2014).

7.2.4 Imbalance of Oxidative Stress-Related Enzymes in PD

The delicate equilibrium between ROS-producing enzymes and antioxidant defenses is disrupted in PD. The dysregulation of enzymatic systems creates a self-perpetuating cycle of oxidative stress, leading to progressive neuronal damage. The interplay between NADPH oxidase, xanthine oxidase, and antioxidant enzymes is a complex dance that underlies the neurodegenerative cascade in PD (Singh et al. 2019).

7.2.5 Modulation of Enzymatic Antioxidant Defenses as a Therapeutic Strategy

7.2.5.1 Unveiling Therapeutic Avenues

Understanding the intricate network of enzymes involved in oxidative stress opens avenues for therapeutic interventions. Modulating enzymatic antioxidant defenses emerges as a promising strategy to mitigate PD progression. Targeting specific enzymes involved in ROS generation or enhancing the activity of antioxidant enzymes presents novel approaches for disease-modifying therapies (Abd Al Moaty et al. 2022).

7.2.5.2 Enhancing SOD Activity

Augmenting SOD activity is a potential therapeutic avenue. Strategies to enhance SOD expression or introduce exogenous SOD may counteract the diminished enzymatic capacity in PD. Gene therapies or pharmacological agents that stimulate SOD expression could hold promise in restoring redox balance (Michiels et al. 1994).

7.2.5.3 Boosting Glutathione Peroxidase and Catalase

Similarly, strategies aimed at bolstering glutathione peroxidase and catalase activities represent attractive therapeutic targets. Selenium supplementation, which is essential for glutathione peroxidase function, or small molecule activators of catalase could enhance the cell's ability to neutralize hydrogen peroxide and mitigate oxidative damage (Kim et al. 2020).

7.2.5.4 Multitarget Approaches

Considering the complexity of oxidative stress in PD, multi-target approaches may hold the key to effective interventions. Combining strategies to reduce ROS production, enhance antioxidant enzyme activity, and promote overall cellular resilience may provide a synergistic therapeutic effect (Grünewald et al. 2019).

7.3 Enzymes and Protein Aggregation

PD is a complex neurodegenerative disorder with a multifactorial etiology. Enzymes are emerging as key players in the pathogenesis of PD, influencing various aspects of the disease, including dopamine metabolism, oxidative stress, protein aggregation, and mitochondrial dysfunction. Dysregulation of these enzymatic processes contributes to the progressive degeneration of dopaminergic neurons in the substantia nigra, leading to the characteristic motor and non-motor symptoms of PD. Understanding the intricate role of enzymes in PD can provide valuable insights into the development of targeted therapeutic approaches aimed at restoring enzyme function and mitigating disease progression. Further research and advancements in enzyme-targeted therapies hold promise for improving the quality of life for individuals living with PD (Maiti et al. 2017).

7.3.1 Protein Aggregation and PD

PD is a progressive neurodegenerative disorder characterized by the accumulation of abnormal protein aggregates in specific regions of the brain, particularly the substantia nigra, leading to the loss of dopaminergic neurons and subsequent motor and non-motor symptoms. These protein aggregates mainly consist of misfolded or aberrant proteins, such as α-synuclein, which form insoluble fibrils and disrupt cellular functions. The pathological hallmark of PD is the formation of Lewy bodies, which are intracellular protein aggregates primarily composed of α-synuclein. Understanding the role of enzymes in protein misfolding and aggregation is crucial in comprehending the pathogenesis of PD and exploring potential therapeutic strategies (Houck et al. 2012).

7.3.2 Role of Enzymes in Protein Misfolding and Aggregation

Protein quality control mechanisms are essential cellular processes that prevent the accumulation of misfolded and aggregated proteins. Enzymes play a critical role in these mechanisms, facilitating proper protein folding, repairing misfolded proteins, and clearing aggregates to maintain cellular homeostasis. Several key enzymes are involved in the prevention and regulation of protein aggregation in cells (Saibil 2013).

7.3.2.1 Chaperone Proteins

Chaperone proteins are a class of enzymes that assist in protein folding, preventing the misfolding and aggregation of nascent polypeptides. They facilitate the correct folding pathways, stabilize intermediate folding states, and prevent protein aggregation by providing a protective environment for vulnerable regions. Heat shock proteins (HSPs) are one of the most well-studied chaperone families implicated in protein quality control. In PD, HSPs have been found to interact with α-synuclein, modulating its aggregation propensity and promoting its degradation, thereby potentially reducing its toxic effects (Thibaudeau and Smith 2019).

7.3.2.2 Ubiquitin-Proteasome System (UPS)

The UPS is a major cellular pathway responsible for degrading short-lived and misfolded proteins. The process involves tagging target proteins with ubiquitin molecules, which mark them for degradation by the proteasome. In PD, dysfunction of the UPS has been linked to the accumulation of misfolded α-synuclein, leading to the formation of Lewy bodies. Mutations in parkin and UCHL1, both key enzymes involved in the UPS, have been associated with familial forms of PD, highlighting the critical role of the UPS in regulating protein aggregation (Malik et al. 2019).

7.3.2.3 Autophagy-Lysosomal Pathway

Autophagy is a lysosome-dependent degradation pathway responsible for clearing aggregated proteins, damaged organelles, and other cellular debris. The process involves the formation of autophagosomes that sequester cytoplasmic cargo and subsequently fuse with lysosomes for degradation. Dysregulation of autophagy has been implicated in the pathogenesis of PD, as impaired clearance of α-synuclein aggregates leads to their accumulation in affected neurons. Enhancing autophagy through pharmacological or genetic means has shown potential in reducing protein aggregation and neurodegeneration in PD models (Cook et al. 2012).

7.3.3 Dysfunction of Enzymes in Protein Quality Control in PD

In PD, various factors can lead to the dysfunction of enzymes involved in protein quality control mechanisms, contributing to the accumulation of misfolded proteins and the development of pathology. Genetic mutations in genes encoding these enzymes, as well as environmental factors and oxidative stress, can compromise their activity and impair their ability to regulate protein aggregation (Pajarillo et al. 2019).

Mutations in genes such as encoding α-synuclein and encoding leucine-rich repeat kinase 2 have been identified in familial forms of PD and are associated with aberrant protein aggregation. Additionally, dysfunction of chaperone proteins, UPS components, and autophagy-related enzymes has been observed in both familial and sporadic cases of PD (Cook et al. 2012).

7.3.4 Therapeutic Approaches Targeting Protein Aggregation Enzymes

Given the prominent role of protein misfolding and aggregation in PD, therapeutic strategies targeting enzymes involved in protein quality control have garnered significant interest. These approaches aim to enhance the clearance of misfolded proteins, prevent their aggregation, and restore cellular proteostasis.

One therapeutic avenue involves the use of pharmacological chaperones that can stabilize the native conformation of misfolded proteins, thereby promoting their proper folding and preventing aggregation. Another strategy focuses on enhancing the activity of chaperone proteins, such as HSPs, to improve their ability to refold misfolded proteins and direct them toward degradation pathways.

Furthermore, the modulation of the UPS and autophagy pathways represents potential therapeutic targets. Small molecules that enhance proteasome activity or promote autophagosome formation could facilitate the clearance of aggregated proteins. Additionally, gene therapy approaches aimed at restoring the function of enzymes with mutations associated with PD could hold promise in reducing protein aggregation (Oliveira et al. 2021).

7.4 Potential Therapeutic Strategies Targeting Enzymes

As researchers delve deeper into the intricate mechanisms underlying PD, there is a growing focus on developing therapeutic strategies that target specific enzymes involved in the pathological processes. One promising avenue is centered around enzymes responsible for the metabolism of alpha-synuclein, the protein implicated in the formation of Lewy bodies—a hallmark of PD (Oliveira et al. 2021).

7.4.1 Alpha-Synuclein Clearance

Enhancing the clearance of abnormal alpha-synuclein aggregates is a key objective for researchers. Enzymes involved in the degradation of alpha-synuclein, such as lysosomal enzymes, have become targets for novel therapeutic interventions. Strategies to boost the activity of these enzymes or prevent the accumulation of alpha-synuclein could potentially slow or halt the progression of PD (Mithaiwala et al. 2021).

7.4.2 Enzyme Modulation

Modulating the activity of specific enzymes implicated in neuroinflammation is another avenue of exploration. Targeting enzymes that regulate the inflammatory response in the brain may attenuate the detrimental effects of neuroinflammation in PD. This approach holds promise for mitigating disease progression and improving overall outcomes for affected individuals (Smith et al. 2012).

7.4.3 Advancements in Enzyme-Targeted Therapies

Recent years have witnessed significant advancements in the development of enzyme-targeted therapies for PD. While many of these approaches are still in the preclinical or early clinical stages, they offer hope for more targeted and effective interventions (Coune et al. 2012).

7.4.3.1 Gene Therapies

The emerging field of gene therapy holds considerable promise for PD. Researchers are exploring the use of viral vectors to deliver therapeutic genes into the brain, aiming to enhance the production of beneficial enzymes or inhibit those contributing to disease progression. Gene therapies have the potential to provide long-lasting effects, reducing the need for frequent administration of medications (Peña-Díaz et al. 2023).

7.4.3.2 Small Molecule Inhibitors

Small molecules that selectively inhibit specific enzymes implicated in PD are being actively investigated. These inhibitors aim to modulate the activity of enzymes associated with alpha-synuclein aggregation, neuroinflammation, and other key pathological processes. The development of highly selective and potent small molecule inhibitors represents a promising strategy for precision medicine in PD treatment (Valera et al. 2016).

7.4.3.3 Immunotherapies

Immunotherapeutic approaches are being explored to target alpha-synuclein and promote the clearance of abnormal protein aggregates. Enzymes involved in the immune response, such as those facilitating the recognition and removal of pathological proteins, are the focus of these innovative therapies. Immunotherapies may offer a disease-modifying strategy by addressing the root cause of PD pathology (Yadav et al. 2022).

7.4.3.4 Protein Degradation Modulators

Strategies to enhance the degradation of abnormal proteins, including alpha-synuclein, are under investigation. Enzymes involved in the UPS, responsible for the targeted degradation of proteins, are potential targets. Modulating these enzymes could facilitate the removal of toxic protein aggregates, offering a novel therapeutic approach.

7.5 Conclusion

To sum up, the complex relationship between enzymes and the pathophysiology of PD highlights the complexity of this neurodegenerative illness. This thorough investigation has illuminated the critical role of enzymes in PD in a number of areas, such as protein aggregation, mitochondrial dysfunction, oxidative stress, and dopamine metabolism. The disruption of these enzymatic pathways contributes to the progressive degeneration of dopaminergic neurons in the substantia nigra, which in turn causes the iconic motor and non-motor symptoms of the disease. The delicate balance between antioxidant defenses and enzymes that produce ROS is highlighted by the role of enzymes in oxidative stress, a major factor in PD. A self-reinforcing loop of oxidative stress is produced by the dysregulation of NADPH oxidase, xanthine oxidase, glutathione peroxidase, SOD, and catalase, which exacerbates neurodegeneration. Comprehending these enzymatic imbalances provides opportunities for treatments; methods aimed at increasing the activity of antioxidant enzymes are emerging as potentially effective techniques.

Furthermore, enzymes are essential for controlling the misfolding and aggregation of proteins, which helps to generate Lewy bodies, a degenerative feature of PD. The autophagy-lysosomal pathway, the UPS, and chaperone proteins are crucial elements of cellular protein quality control systems. When these enzymes are not working properly, misfolded proteins like α-synuclein can build up and cause dementia. By enhancing protein clearance and reestablishing cellular proteostasis, therapeutic approaches that target these enzymes may be able to modify illness.

Creating PD treatments that target specific enzymes is becoming more and more important as research goes on. Developments in protein degradation modulators, immunotherapies, gene treatments, and small molecule inhibitors provide promise for more focused and efficient interventions. These novel strategies may be able to treat the underlying pathologies of PD, which might lead to better results and a higher standard of living for those who have the disease.

In conclusion, deciphering the complex function that enzymes play in PD has advanced our knowledge of the condition and opened the door for novel treatment approaches. The continuous investigation of enzymatic pathways serves as a basis for the creation of tailored and accurate interventions, advancing the prospect of revolutionary therapies for PD.

References

Abd Al Moaty MN, El Ashry ES, Awad LF, Mostafa A, Abu-Serie MM, Teleb M (2022) Harnessing ROS-induced oxidative stress for halting colorectal cancer via thiazolidinedione-based SOD inhibitors. ACS Omega 7(24):21267–21279

Alexander GE (2004) Biology of Parkinson's disease: pathogenesis and pathophysiology of a multisystem neurodegenerative disorder. Dialogues Clin Neurosci 6(3):259–280

Benabid AL (2003) Deep brain stimulation for Parkinson's disease. Curr Opin Neurobiol 13(6):696–706

Berardelli A, Rothwell JC, Thompson PD, Hallett M (2001) Pathophysiology of bradykinesia in Parkinson's disease. Brain 124(11):2131–2146

Checa J, Aran JM (2020) Reactive oxygen species: drivers of physiological and pathological processes. J Inflamm Res 13:1057–1073

Choi J, Horner KA (2019) Dopamine agonists. Cleveland Clinic, Cleveland

Cook C, Stetler C, Petrucelli L (2012) Disruption of protein quality control in Parkinson's disease. Cold Spring Harb Perspect Med 2(5):a009423

Coune PG, Schneider BL, Aebischer P (2012) Parkinson's disease: gene therapies. Cold Spring Harb Perspect Med 2(4):a009431

DeMaagd G, Philip A (2015) Parkinson's disease and its management: part 1: disease entity, risk factors, pathophysiology, clinical presentation, and diagnosis. Pharm Therap 40(8):504

Gómez-Benito M, Granado N, García-Sanz P, Michel A, Dumoulin M, Moratalla R (2020) Modeling Parkinson's disease with the alpha-synuclein protein. Front Pharmacol 11:356

Grünewald A, Kumar KR, Sue CM (2019) New insights into the complex role of mitochondria in Parkinson's disease. Prog Neurobiol 177:73–93

Helmich RC, Hallett M, Deuschl G, Toni I, Bloem BR (2012) Cerebral causes and consequences of Parkinsonian resting tremor: a tale of two circuits? Brain 135(11):3206–3226

Houck SA, Singh S, Cyr DM (2012) Cellular responses to misfolded proteins and protein aggregates. In: Ubiquitin family modifiers and the proteasome: reviews and protocols. Springer, Cham, pp 455–461

Kim TY, Leem E, Lee JM, Kim SR (2020) Control of reactive oxygen species for the prevention of Parkinson's disease: the possible application of flavonoids. Antioxidants 9(7):583

Kouli A, Torsney KM, Kuan WL (2018) Parkinson's disease: etiology, neuropathology, and pathogenesis. Exon Publications, Brisbane, pp 3–26

Lin MT, Beal MF (2006) Mitochondrial dysfunction and oxidative stress in neurodegenerative diseases. Nature 443(7113):787–795

Maiti P, Manna J, Dunbar GL (2017) Current understanding of the molecular mechanisms in Parkinson's disease: targets for potential treatments. Transl Neurodeg 6:1–35

Malik BR, Maddison DC, Smith GA, Peters OM (2019) Autophagic and endo-lysosomal dysfunction in neurodegenerative disease. Mol Brain 12:1–21

Marsh L (2013) Depression and Parkinson's disease: current knowledge. Curr Neurol Neurosci Rep 13:1–9

Menza M, Dobkin RD, Marin H, Bienfait K (2010) Sleep disturbances in Parkinson's disease. Mov Disord 25(1):117–122

Michiels C, Raes M, Toussaint O, Remacle J (1994) Importance of Se-glutathione peroxidase, catalase, and Cu/Zn-SOD for cell survival against oxidative stress. Free Radic Biol Med 17(3):235–248

Mithaiwala MN, Santana-Coelho D, Porter GA, O'Connor JC (2021) Neuroinflammation and the kynurenine pathway in CNS disease: molecular mechanisms and therapeutic implications. Cells 10(6):1548

Nandi A, Yan LJ, Jana CK, Das N (2019) Role of catalase in oxidative stress-and age-associated degenerative diseases. Oxidative Med Cell Longev 2019:9613090

Niedzielska E, Smaga I, Gawlik M, Moniczewski A, Stankowicz P, Pera J, Filip M (2016) Oxidative stress in neurodegenerative diseases. Mol Neurobiol 53:4094–4125

Olanow CW, Stocchi F, Lang A (eds) (2011) Parkinson's disease: non-motor and non-dopaminergic features. Wiley, Hoboken

Oliveira LM, Gasser T, Edwards R, Zweckstetter M, Melki R, Stefanis L, Lashuel HA, Sulzer D, Vekrellis K, Halliday GM, Tomlinson JJ (2021) Alpha-synuclein research: defining strategic moves in the battle against Parkinson's disease. NPJ Parkinson's Dis 7(1):65

Pajarillo E, Rizor A, Lee J, Aschner M, Lee E (2019) The role of posttranslational modifications of α-synuclein and LRRK2 in Parkinson's disease: potential contributions of environmental factors. Biochim Biophys Acta 1865(8):1992–2000

Palakurthi B, Burugupally SP (2019) Postural instability in Parkinson's disease: a review. Brain Sci 9(9):239

Palma JA, Kaufmann H (2018) Treatment of autonomic dysfunction in Parkinson disease and other synucleinopathies. Mov Disord 33(3):372–390

Peña-Díaz S, García-Pardo J, Ventura S (2023) Development of small molecules targeting α-synuclein aggregation: a promising strategy to treat Parkinson's disease. Pharmaceutics 15(3):839

Pratomo IP, Noor DR, Kusmardi K, Rukmana A, Paramita RI, Erlina L, Fadilah F, Gayatri A, Fitriani M, Purnomo TT, Ariane A (2021) Xanthine oxidase-induced inflammatory responses in respiratory epithelial cells: a review in immunopathology of COVID-19. Int J Inflamm 2021:1

Prochazka A, Bennett DJ, Stephens MJ, Patrick SK, Sears-Duru R, Roberts T, Jhamandas JH (1997) Measurement of rigidity in Parkinson's disease. Mov Disord 12(1):24–32

Rahal A, Kumar A, Singh V, Yadav B, Tiwari R, Chakraborty S, Dhama K (2014) Oxidative stress, prooxidants, and antioxidants: the interplay. Biomed Res Int 2014:761264

Rao SS, Hofmann LA, Shakil A (2006) Parkinson's disease: diagnosis and treatment. Am Fam Physician 74(12):2046–2054

Saibil H (2013) Chaperone machines for protein folding, unfolding and disaggregation. Nat Rev Mol Cell Biol 14(10):630–642

Schapira AH (2011) Monoamine oxidase B inhibitors for the treatment of Parkinson's disease: a review of symptomatic and potential disease-modifying effects. CNS Drugs 25:1061–1071

Singh A, Kukreti R, Saso L, Kukreti S (2019) Oxidative stress: a key modulator in neurodegenerative diseases. Molecules 24(8):1583

Smith Y, Wichmann T, Factor SA, DeLong MR (2012) Parkinson's disease therapeutics: new developments and challenges since the introduction of levodopa. Neuropsychopharmacology 37(1):213–246

Thibaudeau TA, Smith DM (2019) A practical review of proteasome pharmacology. Pharmacol Rev 71(2):170–197

Ursini F, Bindoli A (1987) The role of selenium peroxidases in the protection against oxidative damage of membranes. Chem Phys Lipids 44(2-4):255–276

Valera E, Spencer B, Masliah E (2016) Immunotherapeutic approaches targeting amyloid-β, α-synuclein, and tau for the treatment of neurodegenerative disorders. Neurotherapeutics 13:179–189

Wang Q, Liu Y, Zhou J (2015) Neuroinflammation in Parkinson's disease and its potential as therapeutic target. Transl Neurodeg 4:1–9

Weiner WJ, Shulman LM, Lang AE (2013) Parkinson's disease: a complete guide for patients and families. JHU Press, Baltimore

Yadav D, Lee JY, Puranik N, Chauhan PS, Chavda V, Jin JO, Lee PC (2022) Modulating the ubiquitin–proteasome system: a therapeutic strategy for autoimmune diseases. Cells 11(7):1093

Younus H (2018) Therapeutic potentials of superoxide dismutase. Int J Health Sci 12(3):88

Zahoor I, Shafi A, Haq E (2018) Pharmacological treatment of Parkinson's disease. Exon Publications, Brisbane, pp 129–144

Role of Enzymes in Prion Disease: A Molecular Mechanism

8

Sayed Mohammed Firdous, Sourav Pal,
and Subhajit Mandal

Abstract

As an assortment of catastrophic neurodegenerative diseases, prion diseases are distinguished by the buildup of misfolded prion proteins in the brain and spinal cord. Due to exposure to these misfolded proteins, the normal cellular isoform prion protein (PrP-C) is converted into the disease-associated variant scrapie prion protein (PrP-Sc). The importance of enzymes in the etiology of prion disease has recently been elucidated by several research. During the initial stages of prion infection, host-encoded enzymes, such as protein disulfide isomerase, facilitate the conversion of PrP-C to PrP-Sc by promoting the formation of disulfide bonds. This conversion triggers a chain reaction, accumulating PrP-Sc aggregates that disrupt cellular functions and trigger neurotoxicity. Proteolytic enzymes, including cathepsins and matrix metalloproteinases, are involved in the progression of prion diseases. They participate in the proteolytic processing of PrP-Sc, generating shorter fragments that may serve as seeds for prion propagation. These enzymes can also facilitate the release and intercellular spread of prions by cleaving PrP-Sc from the cell surface or within endocytic compartments. Enzymatic activities are also implicated in the clearance of prions from the brain. Misfolded proteins, including PrP-Sc, undergo degradation through the ubiquitin-proteasome system and the autophagy pathways. Enzymes involved

S. M. Firdous (✉)
Department of Pharmacology, Calcutta Institute of Pharmaceutical Technology & AHS, Howrah, West Bengal, India

S. Pal
Department of Pharmacology, Seacom Pharmacy College, Jaladhulagori, Sankrail, West Bengal, India

S. Mandal
Department of Pharmaceutical Sciences, Faculty of Science and Engineering, Dibrugarh University, Dibrugarh, India

B. Mathew, D. G. T. Parambi (eds.), *Enzymes in Neurodegenerative Disorders*,
https://doi.org/10.1007/978-981-97-6822-6_8

in these pathways, such as the E3 ubiquitin ligases and lysosomal hydrolases, are critical in targeting PrP-Sc for degradation and reducing its accumulation. Emerging therapeutic strategies for prion diseases aim to modulate the activity of enzymes involved in prion propagation and clearance. Novel drug candidates targeting specific enzymes, such as protein disulfide isomerase inhibitors or proteasomal degradation activators, have shown promising results in preclinical studies and hold potential for therapeutic interventions. In conclusion, enzymes play a multifaceted role in prion diseases, from facilitating the conversion of PrP-C to PrP-Sc to promoting prion propagation and clearance. Understanding the intricate interplay between enzymes and prion proteins is crucial for unraveling the molecular mechanisms underlying prion diseases and developing effective therapeutic approaches to combat these devastating disorders.

Keywords

Prion misfolding · PrP-Sc accumulation · Proteases clearance · Ubiquitin-proteasome system · Chaperone dysfunction · Oxidative assault

8.1 Introduction

A set of neurodegenerative disorders known as prion diseases are characterized by an accumulation of misfolded proteins, or prions, in the central nervous system (CNS) (Hu et al. 2008). These illnesses include fatal familial insomnia (FFI), variant Creutzfeldt-Jakob disease (CJD), Gerstmann-Sträussler-Scheinker syndrome (GSS), and CJD (Hu et al. 2008). Prion diseases can develop spontaneously, due to a person's genetic makeup, or through transmission. The transformation of the benign cellular prion protein (PrP-C) into the pathogenic isoform (PrP-Sc) is the defining feature of prion disorders (Watts and Bourkas 2018). High-sheet content and resistance to proteolytic degradation characterize the misfolded PrPSc. PrP-Sc buildup damages neurons and ultimately results in severe neurological dysfunction and, in most cases, death (Hughes and Halliday 2017). Despite substantial advancements in our knowledge of the pathology and etiology of prion disorders, there are still no proven treatment approaches. By controlling numerous biological processes, such as protein folding, breakdown, and clearance, enzymes are essential for maintaining cellular homeostasis (Sies and Jones 2020). Enzyme involvement in prion disorders and their potential as therapeutic targets have been the focus of recent research. Proteases, enzymes in charge of breaking down proteins, are a crucial area of research (Malgieri and Grasso 2014). Proteases play a critical function in preserving protein homeostasis by cleaving aberrant or misfolded proteins. Numerous studies have emphasized the role of proteases in the breakdown of prion proteins, including the ubiquitin-proteasome system (UPS) and the autophagy-lysosome pathway (Jellinger 2010). The course of the disease may be aided by defective PrP-Sc clearance brought on by the dysregulation of these proteolytic systems (Azam et al. 2021). Investigating how these proteases interact with prion proteins may reveal possible treatment targets for accelerating protein degradation and lowering prion load. Chaperones are an additional class of enzymes that play a role in protein folding, quality control, and proteases. Protein folding is aided by molecular chaperones, which also work to stop misfolding and encourage the refolding or destruction of misfolded proteins (Kevei et al. 2017). The pathophysiology of prion disorders has been linked to disturbances in chaperone activity. The variables affecting prion protein misfolding and aggregation can be better understood by understanding the molecular interactions between prion proteins and chaperones (Ajmal 2023).

Other enzymes, such as kinases and phosphatases, have also recently been implicated in the pathophysiology of prion disease (Michel and Bakovic 2007). Protein phosphorylation, a crucial post-translational alteration involved in numerous physiological functions, is regulated by kinases and phosphatases (Arc et al. 2011). Alterations in phosphorylation patterns in prion illnesses raise the possibility that kinase and phosphatase dysregulation contribute to the metabolism of prion proteins and their neurotoxicity (Amro et al. 2021).

Investigating the molecular processes through which enzymes in prion illnesses function offers a viable route for comprehending disease pathophysiology and creating efficient therapy approaches. Among the crucial enzymes involved in prion protein metabolism and neurotoxicity are proteases, chaperones, kinases, and phosphatases (Dar et al. 2020). New therapeutic strategies intended to slow the progression of prion disease may be made possible by clarifying their functions and identifying suitable targets for intervention. We can attempt to understand the complexity of these tragic neurodegenerative illnesses and provide novel therapeutic options by investigating the complicated connections between enzymes and prion proteins.

8.2 Prion and Redox Imbalance

Neuroinflammatory redox assault is frequently associated with the pathogenic emergence and advancement of prion diseases in animal models and humans (Ransohoff 2016). This neurodegenerative inflammation primarily affects microglial cells, which undergo a phenotypic transition to a "activated" M1 state, resulting in the increased expression of proinflammatory cytokines and increased oxidative/nitrosative assault mechanisms to combat potential pathogens (Bayarsaikhan et al. 2022). Although microglial-derived neuroinflammation is commonly thought of as a process that causes neuronal mortality, particularly in extreme oxidative assault, its function in prion disorders is not as readily apparent (Khan et al. 2022). Microglia may be involved in prion-induced neurodegeneration since they show typical activation close to spongiform lesions and exhibit this behavior before neuronal loss. The nicotinamide adenine dinucleotide phosphate (NADPH) oxidase enzyme, which is highly expressed in activated microglia and produces superoxide radicals and oxidative assault in nearby cells, may exacerbate the neurodegenerative traits seen in prion disorders (Peggion et al. 2020). The antioxidant enzyme superoxide dismutase (SOD) also exhibits decreased expression in the early stages of prion disease, exacerbating the oxidative assault brought on by NADPH oxidase (Prasad and Bondy 2018).

Importantly, microglial activation and proliferation also significantly contribute to effectively removing prions. Pharmacologically induced decreases in microglia result in increased prion deposition and neurotoxicity (Peggion et al. 2020). Additionally, depending on the prion strain and the species infected, the cytokine profile after prion infection differs. In many rodent models of the disease and human cases of CJD, some anti-inflammatory cytokines become more prevalent (Carroll et al. 2015; Pérez Gómez et al. 2022).

While prion-induced oxidative assault is largely caused by activated microglia and up-regulated respiratory burst enzymes like NADPH oxidase, mitochondrial malfunction also plays a significant role in redox dysfunction in prion disorders (Milhavet et al. 2000; Guo et al. 2003). Early in the pathophysiology of prion disorders, mitochondrial oxidation takes place and adds to oxidative assault. Numerous studies using various prion disease models have shown that the majority of differentially expressed proteins after infection are involved in energy metabolism, including tricarboxylic acid cycle proteins (including succinate dehydrogenase, aconitase 2, and malate dehydrogenase), oxidative phosphorylation proteins (such as ATP synthase), and mitochondrial membrane proteins (like mitoflin) (Arnould et al. 2021; Gasperini et al. 2016; Bourgognon et al. 2018). Along with pro-inflammatory microglial activation and diminished primary antioxidant protection from SOD, this extensive mitochondrial malfunction plays a key role in redox dysfunction. It is most likely the primary cause of oxidative assault in prion diseases (Spiers et al. 2023).

8.3 Prion and Nitrosative Assault

The buildup of prions, which are proteins that have become misfolded, in the brain is the defining feature of a series of neurodegenerative disorders known collectively as transmissible spongiform encephalopathies (Ascari et al. 2020). Prions are unique, infectious agents that lack genetic material (DNA or RNA) but can self-propagate by inducing a conformational change in normal PrP-C, converting them into an abnormal PrP-Sc (Caughey and Kraus 2019; Yuan et al. 2014). The misfolded PrP-Sc is highly stable, resistant to proteolytic degradation, and has a propensity to aggregate, forming insoluble fibrils and amyloid plaques (Hughes and Halliday 2017). Nitrosative assault, on the other hand, refers to the dysregulation of nitric oxide (NO) signaling and the excessive production of reactive nitrogen species (RNS) within cells (Yoon et al. 2021). Nitric oxide is a highly reactive molecule produced by various enzymes, including neuronal nitric oxide synthase (nNOS), inducible nitric oxide synthase (iNOS), and endothelial nitric oxide synthase (eNOS) (Begara-Morales et al. 2018; Lundberg and Weitzberg 2022). Under normal physiological conditions, NO plays crucial roles in neurotransmission, immune response, and vascular regulation (Lundberg and Weitzberg 2022). However, excessive or uncontrolled NO production can lead to nitrosative assault, which is implicated in various pathological conditions, including neurodegenerative diseases such as prion diseases (Lundberg and Weitzberg 2022). The elevation of nitric oxide levels in the CNS has been associated with prion disease pathogenesis (Bourgognon et al. 2021). Studies have shown that prion infection or induction of neuroinflammation through lipopolysaccharide treatment can up-regulate iNOS, the enzyme responsible for producing NO in response to inflammatory stimuli (Carroll and Chesebro 2019; Justo and Suemoto 2022). This up-regulation of iNOS leads to increased NO production, which can have detrimental effects on neuronal cells (Justo and Suemoto 2022). Excessive NO can react with superoxide radicals to form highly reactive peroxynitrite ($ONOO^-$), a potent oxidant that can cause oxidative assault to cellular components, including proteins, lipids, and DNA (Islam et al. 2015).

In prion diseases, elevated NO levels and nitrosative assault can impact disease progression. First, studies have demonstrated that increased NO production can up-regulate the expression of the prion protein itself. Lipopolysaccharide-induced inflammatory response and treatment with NO donors have been shown to enhance prion protein mRNA and protein levels through signaling pathways involving methyl ethyl ketone and p38 mitogen-activated protein kinase (Hiller et al. 2021). This suggests a potential positive feedback loop between NO production and prion protein expression, further promoting disease pathology. Furthermore, the excessive nitrosative assault in prion diseases can contribute to neuronal damage and neuroinflammation. Nitrosative assault can lead to protein nitration, which modifies protein structures and functions (Brazier et al. 2014). Nitration of key proteins involved in neuronal signaling and cell survival pathways can disrupt their normal functions and contribute to neuronal dysfunction and cell death (Nakamura et al. 2021). Additionally, nitrosative assault can trigger inflammation and activate microglial cells, the immune cells of the CNS, leading to the release of proinflammatory cytokines and further amplifying neuroinflammation (Islam et al. 2015). Mitochondrial dysfunction is another consequence of nitrosative assault in prion diseases. Mitochondria are highly susceptible to oxidative and nitrosative damage due to their high metabolic activity and exposure to reactive species (Butterfield and Boyd-Kimball 2020). Excessive NO production can directly impair mitochondrial function, leading to energy deficits, increased production of ROS, and compromised cellular homeostasis (Wang et al. 2021). This mitochondrial dysfunction further contributes to prion diseases' overall redox imbalance and neurodegenerative processes (Zhu et al. 2018). In prion disorders, misfolded prion proteins build up, and in the nitrosative assault, nitric oxide signaling is disrupted, and there is an overabundance of RNS (Spiers et al. 2023; Zhu et al. 2018). Nitrosative assault, resulting from increased NO production, can exacerbate prion disease pathology through multiple mechanisms, including up-regulation of the prion protein, neuronal damage via protein nitration, neuroinflammation, and mitochondrial dysfunction (Zhu et al. 2018). Understanding the intricate interplay between prion pathogenesis and nitrosative assault may provide valuable insights into the underlying mechanisms of disease progression and potentially uncover new therapeutic targets for intervention.

8.4 Human Sporadic CFJ Disease

About 85% of all human instances of prion disease are attributable to sporadic Creutzfeldt-Jakob disease (sCJD). The brain's aberrant buildup of prion proteins is the hallmark of this extremely rare and ultimately fatal neurodegenerative disease (Sitammagari and Masood 2018).

8.4.1 Epidemiology

sCJD occurs worldwide, with an estimated annual incidence of 1–2 cases per million people. It primarily affects individuals in their 60s or older, although younger individuals have reported cases. The disease occurs sporadically, meaning there is no known genetic or environmental cause in most cases. However, rare familial forms of CJD do exist, which are associated with specific genetic mutations (Kosorinova et al. 2021; Hara and Sakaguchi 2021).

8.4.2 Clinical Presentation

The clinical presentation of sCJD is variable but typically manifests as rapidly progressive dementia. Patients may initially experience subtle cognitive and behavioral changes, followed by the development of neurological symptoms such as muscle stiffness, myoclonus (involuntary muscle jerks), coordination problems, and visual disturbances. As the disease progresses, patients may experience profound dementia, and severe motor impairment, and eventually become bedridden and non-responsive (Kwon and Kwon 2019; Xu et al. 2019).

8.4.3 Pathology

Accumulation of PrP-Sc (a misfolded prion protein) and the development of amyloid plaques are two pathogenic features of sCJD. PrP-Sc is an unstable isoform of PrP-C generated by a conformational shift from the normal PrP-C. Neuronal death, gliosis (the overproduction of astrocytes), and spongiform degeneration are all symptoms of PrPSc accumulation in the brain (DeArmond and Bajsarowicz 2010; Thomzig et al. 2007; Taraboulos et al. 1992).

8.4.4 Diagnosis

The diagnosis of sCJD is challenging and often requires a combination of clinical evaluation, neuroimaging, and laboratory tests. Clinical criteria, such as rapidly progressive dementia and characteristic neurological symptoms, aid in the initial diagnosis (Hermann et al. 2021; Chen et al. 2023). Neuroimaging techniques, such as magnetic resonance imaging, may reveal specific patterns of brain atrophy or hyperintense signal abnormalities (Watson et al. 2022). The detection of abnormal prion protein in cerebrospinal fluid (CSF) or brain biopsy samples through techniques like protein misfolding cyclic amplification or immunohistochemistry, can provide supportive evidence for the diagnosis (Hermann et al. 2021; Chen et al. 2023).

8.4.5 Treatment and Prognosis

Currently, there is no cure for sCJD, and the treatment is primarily supportive to manage symptoms and ensure patient comfort. Due to its rapid and relentless progression, sCJD is uniformly fatal, with a median survival time of 4–6 months from the onset of symptoms. Palliative care is essential to address the disease's physical, cognitive, and psychological aspects and support patients and their families (Kwon and Kwon 2019).

Human sCJD is an infrequent and devastating neurodegenerative disorder characterized by the accumulation of abnormal prion proteins in the brain. It presents rapidly progressive dementia and neurological symptoms, leading to profound disability and death. Early diagnosis is crucial for appropriate management and supportive care (Xu et al. 2019).

8.5 Prion Proteins: Multiple Forms

Prion proteins exist in multiple forms, and their conformational changes play a critical role in developing prion diseases. These diseases are characterized by the accumulation of misfolded prion proteins, which can propagate and induce further misfolding of normal proteins (Sarnataro et al. 2017).

8.5.1 PrP-C

On the exterior of many cell types, including neuronal cells, is a normal, non-pathogenic variant of the prion protein, PrP-C. The gene PRNP codes for this protein. The widespread conservation of PrP-C across species indicates its critical role in these processes (Westergard et al. 2007). Cell adhesion, cell signaling, and neural development are only some of the processes thought to benefit from PrP-C, a protein whose precise physiological role is still unclear (Panes et al. 2021). PrP-C has an alpha-helical primary structure with a glycosylphosphatidylinositol anchor that can be attached to the cell membrane. Its expression is widespread, but the brain and spinal cord have the most (Panes et al. 2021).

PrP-C is fascinating because it can undergo conformational modifications and transform into a misfolded isoform called PrP-Sc (Westergard et al. 2007). Prion disorders are linked to PrP-Sc because this misfolded protein generates amyloid plaques in the brain. In contrast to PrP-Sc, PrP-C does not promote the misfolding of other PrP-C molecules. Prion illness, however, can also involve PrP-C if it is attracted to PrP-Sc aggregates that are already present. PrP-C undergoes a templated conformational shift, becoming PrP-Sc, amplifying and spreading the disease-associated isoform (Ugalde et al. 2016). Research is ongoing into the normal role of PrP-C and its significance in the pathophysiology of prion disease. New therapeutic approaches for prion illnesses may be gleaned from research on the interactions and interactions involved in converting PrP-C to PrP-Sc (Ugalde et al. 2016).

8.5.2 Scrapie Prion Protein (PrP-Sc)

PrP-Sc is a misfolded protein with unique biochemical and structural features originating from the normal PrP-C via a conformational change. In contrast to PrP-C's mostly alpha-helical structure, PrP-Sc adopts a beta-sheet-rich conformation (Wille and Requena 2018; Baral et al. 2019). This misfolded isoform is extremely stable, with strong resistance to proteolytic degradation and the ability to replicate independently. Prion disorders have been linked to PrP-Sc accumulation and aggregation in the brain. Amyloid plaques, unique to prion disorders, occur when PrP-Sc is present. Aggregated and insoluble PrP-Sc molecules make up these plaques (Baral et al. 2019; Chen et al. 2014). PrP-Sc's amazing capacity to cause misfolding in normally folding PrP-C molecules is one of its most interesting features. In a process known as seeded conversion, PrP-Sc can serve as a template to guide PrP-C into the misfolded conformation (Wille and Requena 2018; Chen et al. 2014). This change increases the buildup of PrP-Sc and so aids in the propagation and worsening of prion diseases. Brain regions where PrP-Sc has accumulated have lost neurons, undergone gliosis (the growth of astrocytes), and developed spongiform degeneration, characterized by the presence of minute vacuoles (Baral et al. 2019). Animal prion illnesses like scrapie in sheep and goats, bovine spongiform encephalopathy (BSE) in cattle, and CWD in deer and elk are all strongly linked to PrP-Sc. It has also been linked to CJD in humans, both familial and sporadic, as well as the more common variant Creutzfeldt-Jakob disease (vCJD) (Wille and Requena 2018; Chen et al. 2014). Uncovering the etiology of prion illnesses and creating effective therapeutic methods requires a thorough understanding of the structure, dissemination, and mechanisms behind PrP-Sc production.

8.5.3 Protease-Resistant PrP (PrPres)

PrPres is a specific form of the prion protein that exhibits resistance to proteolytic digestion by enzymes, such as proteinase K. PrPres is a key pathological feature of prion diseases and serves as a diagnostic marker for these neurodegenerative disorders (Concha-Marambio et al. 2014).

PrPres is derived from the misfolded isoform of the prion protein, known as PrP-Sc (Poggiolini et al. 2013; Hedman et al. 2018). During the conversion of the normal PrP-C to PrP-Sc, a conformational change occurs, adopting a beta-sheet-rich structure. This misfolded form of PrP-Sc has a higher resistance to proteases than the normal PrP-C (Hedman et al. 2018). The resistance of PrPres to proteolytic digestion allows for its accumulation and persistence in affected tissues, especially in the CNS. It can be detected and distinguished from the normal PrP-C through biochemical techniques that involve the selective degradation of PrP-C, leaving behind the protease-resistant PrPres (Poggiolini et al. 2013; Hedman et al. 2018). The detection of PrPres in biological samples, such as brain tissues or CSF, is essential to prion disease diagnosis (Concha-Marambio et al. 2014). Various laboratory methods, including Western blotting, immunohistochemistry, and enzyme-linked immunosorbent assays, have been developed to identify and quantify PrPres in these samples. These techniques contribute to the accurate and early detection of prion diseases. The presence of PrPres in affected tissues correlates with the pathological changes observed in prion diseases (Concha-Marambio et al. 2014; Hedman et al. 2018). It contributes to the formation of amyloid plaques and aggregates, leading to neuronal damage, spongiform degeneration, and the characteristic clinical manifestations of these disorders. The mechanisms involved in converting PrP-C to PrP-Sc and accumulating PrPres are crucial for unraveling the pathogenesis of prion diseases (Poggiolini et al. 2013).

8.5.4 Detergent-Insoluble Prion Protein (PrP-D)

PrP-D denotes a variant of the prion protein that is resistant to detergent solubilization. PrP-D is highly relevant to the pathophysiology of prion diseases, neurodegenerative conditions (Ashok and Hegde 2009). PrP-Sc is the misfolded isoform of the prion protein from which PrP-D is created (Shaked et al. 1999). In contrast to the primarily helical structure of typical PrP-C, PrP-Sc adopts a unique conformation characterized by a sheet-rich structure (Harris 1999). Prion disorders are characterized by the buildup of PrP-Sc in afflicted tissues. Due to its changed conformation and metabolic characteristics, PrP-Sc is insoluble in detergents (Harris 1999; Nakaoke et al. 2000). PrP-Sc is insoluble and does not dissolve in water when subjected to conventional detergent-based solubilization techniques (Ashok and Hegde 2009). Its insolubility distinguishes PrP-D. Biochemical methods, such as Western blotting or immunohistochemistry, can detect and quantify detergent-insoluble PrP-D (Shaked et al. 1999; Harris 1999). Detergent extraction of brain tissue or other damaged samples is commonly used, and the insoluble fraction is then analyzed for PrP-D presence. The pathological alterations seen in prion illnesses are linked to the buildup of PrP-D in the brain. It has a role in the neuropathological hallmarks of prion disorders, such as the development of amyloid plaques, the accumulation of aggregates, and spongiform degeneration (Nakaoke et al. 2000). Detection and characterization of PrP-D in affected tissues provide important diagnostic information for prion diseases (Ashok and Hegde 2009). Furthermore,

Table 8.1 Isoforms of prion proteins and associated pathological conditions

S. No.	Isoforms of prion proteins	Genetic mutation involved in the process	Possible associated neurodegenerative diseases	References
1.	PrP-C (cellular prion protein)	E200K, D178N, V210I, P102L	Prion disease, Gerstmann-Sträussler-Scheinker syndrome	Villar-Piqué et al. (2019), Kim et al. (2018), Di Fede et al. (2019)
2.	PrP-Sc (scrapie prion protein)	Prion protein gene (PRNP)	Prion disease, Gerstmann-Sträussler-Scheinker syndrome, fatal familial insomnia, Kuru disease	Villar-Piqué et al. (2019), Appleby et al. (2022), Takada et al. (2017)
3.	Protease-resistant PrP (PrPres)	Prion protein gene (PRNP)	Creutzfeldt-Jakob disease, Gerstmann-Sträussler-Scheinker syndrome, fatal familial insomnia	Villar-Piqué et al. (2019), Di Fede et al. (2019), Takada et al. (2017)
4.	PrP-D (detergent-insoluble prion protein)	E200K, D178N, V210I, P102L, A117V, F198S, V180I	Prion disease, Creutzfeldt-Jakob disease, fatal familial insomnia, Gerstmann-Sträussler-Scheinker syndrome	Villar-Piqué et al. (2019), Kim et al. (2018), Appleby et al. (2022)

understanding the biochemical and biophysical properties of PrP-D contributes to developing diagnostic assays and identifying potential therapeutic targets aimed at disrupting its formation and mitigating the progression of prion diseases (Nakaoke et al. 2000). PrP-D, as a detergent-insoluble form of the prion protein, represents a distinctive pathological feature of prion diseases and is a diagnostic indicator of these devastating neurodegenerative disorders (Shaked et al. 1999). Depending on the misfolded prion protein's shape and its biological features, prion disorders can manifest in a variety of forms and strains (Ashok and Hegde 2009; Harris 1999). The resistance to proteinase K, aggregation patterns, and disease phenotypes are only a few of the biochemical traits that help differentiate strains from one another (Shaked et al. 1999). The complexity and variation seen in prion illnesses can be attributed in part to the existence of different forms and strains of the prion protein (Shaked et al. 1999; Nakaoke et al. 2000) (Table 8.1).

8.6 Invasion and Spread of Prions: Alimentary Tract, Scarification of Gums, Skin, and Conjunctiva

The invasion and spread of prions, the infectious agents responsible for prion diseases, can occur through various routes, including the alimentary tract (digestive system), scarification of gums, skin, and conjunctiva (the membrane that covers the front of the eye and lines the eyelids). These entry routes facilitate the transmission and dissemination of prions within the body (Gough and Maddison 2010; Weissmann et al. 2002).

8.6.1 Alimentary Tract (Digestive System)

Prions can enter the body by ingesting contaminated material, primarily in the form of prion-infected tissues or food products derived from infected animals (Lee 2023). For example, in BSE or "mad cow disease," prions can be acquired by consuming meat or other products contaminated with nervous system tissues, such as the brain or spinal cord (Olech 2023). Upon ingestion, prions can resist degradation in the digestive system, cross the intestinal barrier to reach lymphoid tissues, and eventually spread to the CNS (Lee 2023).

8.6.2 Scarification of Gums

In certain circumstances, prions can be transmitted through the scarification (abrasion or cuts) of gums (Porter 2003). This transmission route is more relevant in medical procedures, such as contaminated surgical instruments or dental equipment, that may come into contact with infected tissues or fluids (Jayanthi et al. 2013). The entry of prions through scarified gums provides a direct pathway for the agent to access the bloodstream and potentially reach the CNS (Jayanthi et al. 2013).

8.6.3 Skin

Prions can also enter the body through breaks in the skin, such as cuts, wounds, or lesions. This transmission mode is relatively rare but can occur when prion-contaminated materials come into contact with compromised skin barriers (Wang et al. 2019). Occupational exposure to prion-infected tissues, as observed in laboratory or healthcare settings, is a potential risk for skin transmission. However, the efficiency of prion transmission through intact skin is considered to be very low (Wang et al. 2019).

8.6.4 Conjunctiva

The conjunctiva, a mucous membrane, can serve as a potential prion entry point (Singh et al. 2020). Contact with prion-contaminated material, such as tissue extracts or eye instruments contaminated with infectious prions, can lead to transmission through the conjunctiva (Armitage et al. 2009; Orrù et al. 2018). This route has been implicated in transmitting prion diseases in animal experiments and rare cases of human iatrogenic (medical procedure-related) transmission (Armitage et al. 2009).

It is important to note that while prions can enter the body through these routes, their ability to establish infection and cause disease may vary depending on the specific prion strain, the susceptibility of the host species, and other factors (Orrù et al. 2018). Additionally, prion diseases are not considered highly contagious like many infectious diseases, as the transmission generally requires direct exposure to prion-infected materials or tissues (Singh et al. 2020).

Precautionary measures, such as implementing strict food safety regulations, maintaining proper sterilization procedures in healthcare settings, and adhering to safe laboratory practices, are crucial for minimizing the risk of prion transmission through these routes (Singh et al. 2020).

8.7 Molecular and Cellular Pathways Leading to Neurodegeneration

Neurodegenerative diseases such as Alzheimer's disease, Parkinson's disease, and prion diseases share common pathways that contribute to the progressive loss of neuronal structure and function (Majd et al. 2015). One key pathway is protein misfolding and aggregation, where specific proteins in each disease misfold and form insoluble deposits, disrupting cellular function and triggering neurodegeneration (Sweeney et al. 2017). Oxidative assault is another important pathway, caused by ROS that damage cellular components. The balance between ROS and antioxidants is disrupted in neurodegenerative diseases, leading to oxidative assault and neuronal death (Shrestha et al. 2022). Mitochondrial dysfunction is also implicated, as impaired energy production, increased ROS production, and the release of pro-apoptotic factors contribute to neuronal loss (Maurya et al. 2022). Excitotoxicity occurs when glutamate receptors are overactivated, resulting in an influx of calcium ions that can cause mitochondrial dysfunction, oxidative assault, and cell death (Hakiminia et al. 2022). Chronic inflammation also plays a role, with immune cells in the brain becoming chronically activated and releasing pro-inflammatory factors that damage neurons. Impaired autophagy, the cellular process responsible for clearing damaged components, can lead to the accumulation of damaged proteins and organelles in neurodegenerative diseases (Zhang et al. 2023). Finally, dysregulation of apoptosis, a form of programmed cell death, also contributes to the loss of neurons. Understanding these common molecular and cellular pathways provides insights into the mechanisms underlying neurodegenerative diseases and offers potential targets for therapeutic interventions aimed at preserving neuronal structure and function (Hakiminia et al. 2022).

8.7.1 PrPs Neurodegeneration

In prion diseases, such as CJD in humans, the abnormal PrP, or PrP-Sc, induces a conformational change in the normal, PrP-C. This change causes PrP-C to adopt the misfolded, pathogenic form, leading to the accumulation of PrP-Sc in the brain

(Fig. 8.1) (Thellung et al. 2022). Several processes contribute to the progressive damage observed in the brain. PrP-C is a normal, predominantly alpha-helical protein that is anchored to the cell membrane (Lopes et al. 2022). In PrP-Sc, PrP-C takes on a beta-sheet-rich structure, and this is what causes the conformational shift that leads to misfolding. This conformational change can be triggered by various factors, including genetic mutations or exposure to infectious prions (Do Amaral and Cordeiro 2021). Small misfolded PrP-Sc aggregates, called oligomers, are formed as the first stage in the misfolding process and serve as nucleation seeds (Rahman et al. 2022). These protomers serve as models for folding additional PrP-C molecules into the aberrant shape. For other PrP-C molecules to adopt the misfolded conformation, the misfolded PrP-Sc must serve as a template or seed (Sanz-Hernández et al. 2021). PrP-C molecules are enlisted for this template-directed conversion, and their conformation is altered to match that of PrP-Sc. The transformation of PrP-C to PrP-Sc initiates a self-replicating cycle (Rahman et al. 2022). The newly formed PrP-Sc molecules can serve as templates for the conversion of additional PrP-C molecules, leading to the amplification and spread of misfolded PrP-Sc throughout the brain (Thellung et al. 2022). The misfolded PrP-Sc is more stable and resistant to degradation compared to the normal PrP-C. As a result, PrPSc accumulates in the brain, forming insoluble aggregates and amyloid plaques. Misfolded PrPSc has a higher propensity to aggregate and form insoluble deposits in the brain (Zou 2023). These aggregates can accumulate as plaques and amyloid structures, disrupting normal cellular processes. The accumulation of PrP-Sc can lead to proteotoxic assault and impair protein homeostasis in the brain (Zou 2023). This disrupts the normal function of protein quality control systems, such as chaperones and the UPS, responsible for maintaining proper protein folding and degradation (Ji et al. 2019; Amm et al. 2014). The presence of misfolded proteins triggers cellular assault responses, including the unfolded protein response (UPR) and the heat shock response. These responses aim to restore protein homeostasis but can also induce cellular dysfunction and contribute to neuronal damage (Adams et al. 2019). PrP-Sc accumulation and associated cellular assault can lead to synaptic dysfunction. Synapses, which are crucial for neuronal communication, become impaired, leading to disruptions in neuronal signaling and network connectivity (Ma et al. 2019). Prolonged exposure to misfolded PrP-Sc, synaptic dysfunction, and impaired protein homeostasis ultimately result in neuronal injury and death (Vanni et al. 2020). This leads to the progressive neurodegeneration seen in prion

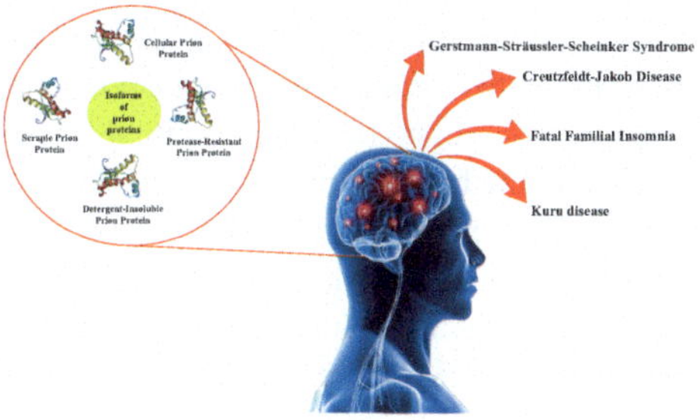

Fig. 8.1 Prion protein-induced pathogenesis

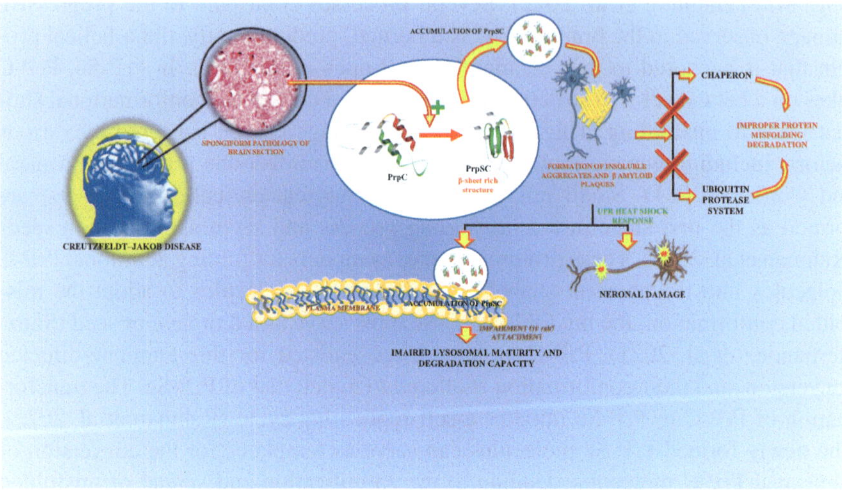

Fig. 8.2 Prp-Sc accumulation and neuronal damage

diseases. Misfolded PrP-Sc and neurodegeneration trigger an immune response in the brain, involving the activation of microglia and astrocytes (Ma et al. 2019). These immune cells release pro-inflammatory molecules, such as cytokines and ROS, which can further contribute to neuronal assault. Prions have the unique ability to convert normal PrP-C into the pathogenic PrP-Sc conformation (Zou 2023). This conversion process allows prions to spread from neuron to neuron, propagating the disease throughout the brain and contributing to the progressive neurodegeneration (Yoshida and Hasegawa 2022) (Fig. 8.2).

8.7.2 Increase Activity of Cu, Zn, SOD, gpx Neuron Expressing Higher Level of Cellular Prp Against Oxidative Assault

Increased activity of copper-zinc (Cu, Zn) SOD and glutathione (GSH) peroxidase (GPx) in neurons expressing higher levels of PrP-C can help protect against oxidative assault (Yusuf et al. 2018). Here is how these mechanisms work:

Neurons expressing higher levels of PrP-c may exhibit increased Cu, Zn SOD activity as a response to oxidative assault or other cellular signals (Zeng et al. 2015). PrP-c expression levels can be influenced by various factors, including genetic and environmental factors. This enzyme converts superoxide radicals into hydrogen peroxide, which can be further detoxified by other antioxidant enzymes. PrP-c expression levels can be influenced by various factors, including genetic and environmental factors (Westergard et al. 2007). During oxidative assault, ROS are generated, including superoxide radicals ($O_2^{\bullet}-$). These superoxide radicals can be produced as byproducts of various cellular processes, including mitochondrial respiration and enzymatic reactions (Peralta and Eyestone 2009). Cu, Zn SOD is located in the cytoplasm and mitochondria of cells, including neurons. When superoxide radicals are generated, Cu, Zn SOD is activated and binds to the superoxide radicals (Yusuf et al. 2018). Cu, Zn SOD facilitates the conversion of superoxide radicals into hydrogen peroxide (H_2O_2) through a dismutation reaction (Westergard et al. 2007). This reaction involves the transfer of electrons within the enzyme's active site, leading to the transformation of $O_2^{\bullet}-$ into H_2O_2. Hydrogen peroxide generated by Cu, Zn SOD can be further detoxified by other antioxidant enzymes, such as catalase and GSH peroxidase (Yusuf et al. 2018). These enzymes convert H_2O_2 into water and molecular oxygen, preventing the accumulation of toxic levels of hydrogen peroxide (Vanni et al. 2020).

The molecular cascade of increased GSH peroxidase (GPx) activity in neurons expressing higher levels of PrP-C against oxidative assault involves the following steps: Neurons with higher levels of PrP-C may exhibit increased GPx activity as a response to oxidative assault or other cellular signals (Panes et al. 2021). PrP-c expression can be influenced by various factors, including genetic and environmental factors (Zeng et al. 2015). During the oxidative assault, ROS, such as H_2O_2 and lipid hydroperoxides, are generated within cells. These ROS can cause oxidative damage to cellular components if not properly controlled (Pisoschi et al. 2021). GPx is an antioxidant enzyme in the cytoplasm and other cellular compartments, including mitochondria (Zhu et al. 2018). It is responsible for reducing hydrogen peroxide and lipid hydroperoxides to their corresponding alcohols using reduced GSH as a cofactor that utilizes GSH to detoxify hydrogen peroxide and lipid hydroperoxides (Weaver and Skouta 2022). In the process, GPx converts GSH to its oxidized form (GSSG). However, GSSG can be recycled back to GSH through the action of other enzymes, such as GSH reductase, ensuring the availability of GSH for continuous GPx activity (Hasanuzzaman et al. 2019) reduces hydrogen peroxide to water and lipid hydroperoxides to their corresponding alcohols. This enzymatic reaction protects cells from the damaging effects of ROS and prevents oxidative assault-induced cellular injury (Westergard et al. 2007).

8.7.3 Protein Disulfide Isomerase (PDI)

PDI is an enzyme involved in protein folding and the formation of disulfide bonds within proteins. In the context of prion diseases, PDI has been implicated in the misfolding and aggregation of the PrP (Otero et al. 2021). Here is a generalized overview of the potential molecular cascade of PDI in prion disease: The normal PrP-C can undergo misfolding, adopting a pathogenic conformation known as PrP-Sc (Gandhi et al. 2019). This misfolding event may be triggered by genetic mutations, exposure to infectious prions, or other factors (Ajmal 2023). During the misfolding process, PrP may interact with PDI. PDI acts as a chaperone and participates in protein folding by catalyzing disulfide bond formation, reduction, and rearrangement (Kanemura et al. 2020). PDI assists in correctly forming disulfide bonds within proteins, contributing to their proper folding and stability (Gandhi et al. 2019). In the case of PrP, the misfolded PrP-Sc isoform is thought to contain incorrect disulfide bond arrangements compared to the normal PrP-C isoform (Spagnolli et al. 2019). The PDI pathway has been hypothesized to contribute to PrP-Sc misfolding and aggregation. Incorrect creation of disulfide bonds in PrP during misfolding has been linked to PDI, which may enhance the conversion of PrP-C to PrP-Sc (Watts et al. 2009). PDI may contribute to the aggregation of misfolded PrP molecules. It has been observed that PDI can interact with misfolded proteins, including PrP-Sc, and participate in the formation of protein aggregates or amyloid fibrils (Nakamura et al. 2021). The involvement of PDI in the misfolding and aggregation of PrP-Sc can disrupt proteostasis, the balance of protein folding, assembly, and degradation within cells (Kraus et al. 2017). The accumulation of misfolded and aggregated PrP-Sc can overwhelm cellular protein quality control systems, leading to proteotoxic assault and subsequent neuronal dysfunction (Lambert et al. 2021).

8.7.4 Cathepsin, Matrix Metalloproteinase (MMP; Generate Short Fragment)

Pathogenesis of prion disorders has been attributed to proteolytic enzymes such as cathepsins and MMPs. In experimental models, cathepsins, such as cathepsin B (CatB), are directly injected into the brain or CNS to induce prion disease (Llorens et al. 2017). This injection bypasses the normal prion propagation routes seen in naturally occurring prion diseases. In prion diseases, including CJD, there is evidence of increased activity of cathepsins, particularly cathepsin B (CatB), in affected brain regions (Huarcaya and Zunke 2023). CatB can degrade both the normal PrP-C and the PrP-Sc. Cathepsins within the lysosomes can target PrP-C or PrP-Sc for degradation, contributing to their turnover and clearance (Chassefeyre et al. 2021). Cathepsins, particularly CatB, have been shown to cleave PrP-Sc, generating shorter fragments. These fragments, known as proteinase K-resistant core fragments, exhibit increased aggregation propensity and enhanced cytotoxicity compared to full-length PrP-Sc (Castle et al. 2023). The cleavage of PrP-Sc by cathepsins generates shorter, aggregation-prone fragments that can accumulate and contribute to the pathogenesis of prion diseases (Thüne et al. 2019). These fragments can interact with neuronal cells, disrupt cellular functions, and induce neurotoxicity (Yadavalli et al. 2004). The accumulation of cathepsin-generated PrP-Sc fragments can trigger neuroinflammation. The interaction between these fragments and microglial cells in the brain leads to the release of pro-inflammatory cytokines, chemokines, and ROS, further exacerbating neurotoxicity (Altmeppen et al. 2012). Cathepsin-mediated degradation and cleavage of PrP-Sc may affect the propagation of prions. The generation of shorter PrP-Sc fragments could potentially contribute to the spread of prions within the CNS (Polyakova et al. 2009).

During prion disease progression, there is evidence of increased expression and activity of MMPs, specifically MMP-2 and MMP-9, in affected brain regions (Cabral-Pacheco et al. 2020). Upregulation of metalloproteinases has been linked to prion disorders, but the precise causes and mechanisms behind this phenomenon remain unclear. The collapse of the blood-brain barrier (BBB) has been linked to

MMPs, specifically MMP-9, in prion disorders (Cabral-Pacheco et al. 2020). Increased MMP activity can lead to the degradation of components of the BBB, such as tight junction proteins and extracellular matrix molecules, compromising the integrity of the BBB (Rempe et al. 2016). The disruption of the BBB allows for the infiltration of immune cells, including lymphocytes and macrophages, into the CNS (Vafadari et al. 2016). The presence of these immune cells contributes to neuroinflammatory responses in prion diseases. MMP-mediated BBB disruption and immune cell infiltration trigger neuroinflammation (Rempe et al. 2016). The activated immune cells release pro-inflammatory cytokines, chemokines, and reactive oxygen species, further exacerbating neurotoxicity and contributing to disease progression (Cabral-Pacheco et al. 2020). MMPs, particularly MMP-9, may facilitate the dissemination of prions within the CNS. The breakdown of the BBB allows prions to spread from the initial infection or accumulation site to other brain regions, promoting disease progression (Vafadari et al. 2016). MMPs have been suggested to influence prion protein aggregation and accumulation. They may interact with prion proteins, promote the conversion of normal PrP-C to the PrP-Sc, and modulate the formation of prion aggregates (Cabral-Pacheco et al. 2020).

8.7.5 UPS

The UPS is a cellular breakdown system for undesired and improperly folded proteins. The UPS is essential for eliminating aberrant, misfolded PrP-Sc in the context of prion disorders (Lin et al. 2013). A possible molecular chain reaction involving UPS in prion disease is summarized as follows: Aggregation of misfolded PrP-Sc in the brain is a hallmark of prion ailments (Baiardi et al. 2023). The abnormal PrP-Sc isoform is resistant to degradation and tends to aggregate, leading to the formation of insoluble protein aggregates and amyloid plaques. In the UPS pathway, proteins targeted for degradation are covalently labeled with ubiquitin molecules (Lin et al. 2013). Ubiquitin ligases recognize and attach ubiquitin molecules to specific sites on misfolded PrP-Sc, marking them for degradation (Baiardi et al. 2023). Ubiquitinated PrP-Sc is recognized by the 26S proteasome, a large protein complex responsible for the degradation of tagged proteins. The proteasome unfolds the tagged PrP-Sc and hydrolyzes it into smaller peptide fragments (Laskowska et al. 2019). The proteasomal degradation of ubiquitinated PrP-Sc fragments facilitates the clearance of misfolded PrP-Sc from the cell. This clearance process helps prevent the accumulation and propagation of PrP-Sc, reducing the burden of pathological prion aggregates (Marrero-Winkens et al. 2020). The UPS is an essential component of cellular proteostasis, maintaining the balance between protein synthesis, folding, and degradation (Baiardi et al. 2023). Dysfunction or impairment of the UPS can lead to the accumulation of misfolded proteins and proteotoxic assault. Studies have suggested that the UPS may be impaired in prion diseases (Marrero-Winkens et al. 2020). The accumulation of misfolded PrP-Sc can overwhelm the UPS machinery, leading to its dysfunction and decreased ability to clear pathological prion aggregates (Vanni et al. 2020). Accumulation of misfolded proteins and

UPS dysfunction can activate assault response pathways, such as the UPR and heat shock response. These assault responses aim to restore proteostasis and mitigate cellular damage caused by protein misfolding (Baiardi et al. 2023).

8.7.6 E3 Ubiquitin Ligase and Lysosomal Hydrolase

E3 ubiquitin ligases and lysosomal hydrolases play important roles in the degradation and clearance of misfolded proteins, including the abnormal PrP-Sc, in prion diseases (López-Pérez et al. 2020). Here is a summary of the potential molecular cascade involving E3 ubiquitin ligases and lysosomal hydrolases in prion disease: E3 ubiquitin ligases play a critical role in tagging proteins for degradation by the UPS (Yadav et al. 2022). In the context of prion diseases, specific E3 ubiquitin ligases can recognize and bind to misfolded PrP-Sc, marking it for degradation. Upon binding of E3 ubiquitin ligases, misfolded PrP-Sc undergoes ubiquitination, wherein multiple ubiquitin molecules are covalently attached to PrP-Sc (Yagami et al. 2019). Ubiquitin serves as a molecular tag that targets the protein for degradation. Ubiquitinated PrP-Sc is then recognized by the 26S proteasome, a large protein complex responsible for protein degradation (Yadav et al. 2022). The proteasome unfolds the ubiquitinated PrP-Sc and breaks it down into smaller peptide fragments. In addition to the UPS, the lysosomal pathway is another major route for protein degradation (Zakany et al. 2023). In prion diseases, misfolded PrP-Sc can also be targeted for degradation via the lysosomal pathway. Chaperone-mediated autophagy (CMA) involves the direct translocation of proteins into the lysosome for degradation (Liu et al. 2023). In CMA, molecular chaperones recognize the misfolded PrP-Sc and deliver it to the lysosome through specific receptor proteins (Wentink et al. 2019). Once inside the lysosome, misfolded PrP-Sc is degraded by various lysosomal hydrolases, including proteases and other enzymes. These lysosomal hydrolases break down the PrP-Sc into smaller peptides and amino acids (Hutti et al. 2020). The degradation of PrP-Sc by lysosomal hydrolases helps clear the abnormal protein from cells and tissues. This clearance process reduces the burden of pathological prion aggregates and mitigates the spread of prion disease (Panda and Mahapatra 2023). Dysfunction or impairment of the UPS, E3 ubiquitin ligases, or lysosomal pathway can lead to the aggregation of misfolded PrP-Sc and contribute to disease progression. Disruptions in these clearance mechanisms can result in the persistence and propagation of prion aggregates (Alyenbaawi et al. 2020).

8.8 Conclusion

In conclusion, prion diseases are complex neurodegenerative disorders characterized by the accumulation of misfolded prion proteins, particularly the pathological isoform PrP-Sc. Multiple forms of PrP, including PrP-C, PrP-Sc, PrPres, and PrP-D, contribute to the development and progression of these devastating diseases. Redox imbalance plays a significant role in prion pathogenesis, and enzymes such as PDI are involved in the conversion of PrP-C to PrP-Sc, which initiates a chain reaction leading to the accumulation of toxic PrP aggregates. Moreover, the involvement of enzymes such as cathepsins and MMPs in prion propagation highlights their importance in processing PrP-Sc and the generation of prion seeds for further spread. The molecular and cellular pathways leading to neurodegeneration in prion diseases are intricate and involve disrupting normal cellular functions by PrP aggregates. The UPS and autophagy pathways, along with enzymes like E3 ubiquitin ligases and lysosomal hydrolases, play crucial roles in the clearance of misfolded proteins, including PrP-Sc, from the brain. Imbalances in these pathways can contribute to the accumulation of PrP aggregates and subsequent neurodegeneration. Understanding the interplay between enzymes and prion proteins is vital for unraveling the molecular mechanisms underlying prion diseases. Therapeutic interventions targeting these enzymes, such as PDI inhibitors or activators of proteasomal degradation, show promise in preclinical studies and hold potential for developing effective treatments for prion diseases.

References

Adams CJ, Kopp MC, Larburu N, Nowak PR, Ali MM (2019) Structure and molecular mechanism of ER stress signaling by the unfolded protein response signal activator IRE1. Front Mol Biosci 6:11

Ajmal MR (2023) Protein misfolding and aggregation in proteinopathies: causes, mechanism and cellular response. Diseases 11:30

Altmeppen HC, Puig B, Dohler F, Thurm DK, Falker C, Krasemann S, Glatzel MJ (2012) Proteolytic processing of the prion protein in health and disease. Am J Neurodegener Dis 1:15

Alyenbaawi H, Allison WT, Mok S-A (2020) Prion-like propagation mechanisms in tauopathies and traumatic brain injury: challenges and prospects. Biomolecules 10:1487

Amm I, Sommer T, Wolf DH (2014) Protein quality control and elimination of protein waste: the role of the ubiquitin–proteasome system. Biochim Biophys Acta 1843:182–196

Amro Z, Yool AJ, Collins-Praino LE (2021) The potential role of glial cells in driving the prion-like transcellular propagation of tau in tauopathies. Brain Behav Immun Health 14:100242

Appleby BS, Shetty S, Elkasaby M (2022) Genetic aspects of human prion diseases. Front Neurol 13:1003056

Arc E, Galland M, Cueff G, Godin B, Lounifi I, Job D, Rajjou L (2011) Reboot the system thanks to protein post-translational modifications and proteome diversity: how quiescent seeds restart their metabolism to prepare seedling establishment. Proteomics 11:1606–1618

Armitage W, Tullo A, Ironside J (2009) Risk of Creutzfeldt–Jakob disease transmission by ocular surgery and tissue transplantation. Eye 23:1926–1930

Arnould H, Baudouin V, Baudry A, Ribeiro LW, Ardila-Osorio H, Pietri M, Caradeuc C, Soultawi C, Williams D, Alvarez M (2021) Loss of prion protein control of glucose metabolism promotes neurodegeneration in model of prion diseases. PLoS Pathog 17:e1009991

Ascari LM, Rocha SC, Gonçalves PB, Vieira TC, Cordeiro Y (2020) Challenges and advances in antemortem diagnosis of human transmissible spongiform encephalopathies. Biotechnology 8:585896

Ashok A, Hegde RS (2009) Selective processing and metabolism of disease-causing mutant prion proteins. PLoS Pathog 5:e1000479

Azam S, Haque ME, Kim I-S, Choi D-K (2021) Microglial turnover in ageing-related neurodegeneration: therapeutic avenue to intervene in disease progression. Cells 10:150

Baiardi S, Mammana A, Capellari S, Parchi P (2023) Human prion disease: molecular pathogenesis, and possible therapeutic targets and strategies. Expert Opin Ther Targets 27(12):1271–1284

Baral PK, Yin J, Aguzzi A, James MN (2019) Transition of the prion protein from a structured cellular form (PrPC) to the infectious scrapie agent (PrPSc). Protein Sci 28:2055–2063

Bayarsaikhan G, Bayarsaikhan D, Lee J, Lee B (2022) Targeting scavenger receptors in inflammatory disorders and oxidative stress. Antioxidants 11:936

Begara-Morales JC, Chaki M, Valderrama R, Sánchez-Calvo B, Mata-Pérez C, Padilla MN, Corpas FJ, Barroso J (2018) Nitric oxide buffering and conditional nitric oxide release in stress response. J Exp Bot 69:3425–3438

Bourgognon J-M, Spiers JG, Scheiblich H, Antonov A, Bradley SJ, Tobin AB, Steinert JR (2018) Alterations in neuronal metabolism contribute to the pathogenesis of prion disease. Differentiation 25:1408–1425

Bourgognon J-M, Spiers JG, Robinson SW, Scheiblich H, Glynn P, Ortori C, Bradley SJ, Tobin AB, Steinert JR (2021) Inhibition of neuroinflammatory nitric oxide signaling suppresses glycation and prevents neuronal dysfunction in mouse prion disease. Proc Natl Acad Sci USA 118:e2009579118

Brazier MW, Wedd AG, Collins SJ (2014) Antioxidant and metal chelation-based therapies in the treatment of prion disease. Antioxidants 3:288–308

Butterfield DA, Boyd-Kimball D (2020) Mitochondrial oxidative and nitrosative stress and Alzheimer disease. Antioxidants 9:818

Cabral-Pacheco GA, Garza-Veloz I, Castruita-De la Rosa C, Ramirez-Acuna JM, Perez-Romero BA, Guerrero-Rodriguez JF, Martinez-Avila N, Martinez-Fierro ML (2020) The roles of matrix metalloproteinases and their inhibitors in human diseases. Int J Mol Sci 21:9739

Carroll JA, Chesebro BJV (2019) Neuroinflammation, microglia, and cell-association during prion disease. Viruses 11:65

Carroll JA, Striebel JF, Race B, Phillips K, Chesebro BJ (2015) Prion infection of mouse brain reveals multiple new upregulated genes involved in neuroinflammation or signal transduction. J Virol 89:2388–2404

Castle AR, Kang S-G, Eskandari-Sedighi G, Wohlgemuth S, Nguyen M-A, Drucker DJ, Mulvihill EE, Westaway DJ (2023) Beta-endoproteolysis of the cellular prion protein by dipeptidyl peptidase-4 and fibroblast activation protein. Proc Natl Acad Sci USA 120:e2209815120

Caughey B, Kraus A (2019) Transmissibility versus pathogenicity of self-propagating protein aggregates. Viruses 11:1044

Chassefeyre R, Chaiamarit T, Verhelle A, Novak SW, Andrade LR, Leitão AD, Manor U, Encalada SE (2021) Endosomal sorting drives the formation of axonal prion protein endoggresomes. Sci Adv 7:3693

Chen W, van der Kamp MW, Daggett VJ (2014) Structural and dynamic properties of the human prion protein. Biophys J 106:1152–1163

Chen Z, Kong Y, Zhang J, Chu M, Liu L, Xie K, Cui Y, Ye H, Li J, Wang L (2023) Toward an early clinical diagnosis of MM2-type sporadic Creutzfeldt–Jakob disease. Ann Clin Transl Neurol 10(7):1209–1218

Concha-Marambio L, Diaz-Espinoza R, Soto CJ (2014) The extent of protease resistance of misfolded prion protein is highly dependent on the salt concentration. J Biol Chem 289:3073–3079

Dar KB, Bhat AH, Amin S, Reshi BA, Zargar MA, Masood A, Ganie SA (2020) Elucidating critical proteinopathic mechanisms and potential drug targets in neurodegeneration. Neurobiology 40:313–345

DeArmond SJ, Bajsarowicz K (2010) PrPSc accumulation in neuronal plasma membranes links Notch-1 activation to dendritic degeneration in prion diseases. Mol Neurodegener 5:1–13

Di Fede G, Catania M, Atzori C, Moda F, Pasquali C, Indaco A, Grisoli M, Zuffi M, Guaita MC, Testi R (2019) Clinical and neuropathological phenotype associated with the novel V189I mutation in the prion protein gene. Acta Neuropathol Commun 7:1–11

Do Amaral MJ, Cordeiro YJ (2021) Intrinsic disorder and phase transitions: pieces in the puzzling role of the prion protein in health and disease. Science 183:1–43

Gandhi J, Antonelli AC, Afridi A, Vatsia S, Joshi G, Romanov V, Murray IV, Khan SA (2019) Protein misfolding and aggregation in neurodegenerative diseases: a review of pathogeneses, novel detection strategies, and potential therapeutics. Rev Neurosci 30:339–358

Gasperini L, Meneghetti E, Legname G, Benetti F (2016) In absence of the cellular prion protein, alterations in copper metabolism and copper-dependent oxidase activity affect iron distribution. Front Neurosci 10:437

Gough KC, Maddison BC (2010) Prion transmission: prion excretion and occurrence in the environment. Prion 4:275–282

Guo C, Sun L, Chen X, Zhang D (2003) Oxidative stress, mitochondrial damage and neurodegenerative diseases. Neural Regen Res 2013:8

Hakiminia B, Alikiaii B, Khorvash F, Mousavi S (2022) Oxidative stress and mitochondrial dysfunction following traumatic brain injury: from mechanistic view to targeted therapeutic opportunities. Fundam Clin Pharmacol 36:612–662

Hara H, Sakaguchi SJ (2021) Virus infection, genetic mutations, and prion infection in prion protein conversion. Int J Mol Sci 22:12439

Harris DA (1999) Cellular biology of prion diseases. Clin Microbiol Rev 12:429–444

Hasanuzzaman M, Bhuyan MB, Anee TI, Parvin K, Nahar K, Mahmud JA, Fujita M (2019) Regulation of ascorbate-glutathione pathway in mitigating oxidative damage in plants under abiotic stress. Antioxidants 8:384

Hedman C, Otero A, Douet J-Y, Lacroux C, Lugan S, Filali H, Corbière F, Aron N, Badiola JJ, Andréoletti O (2018) Detection of PrPres in peripheral tissue in pigs with clinical disease induced by intracerebral challenge with sheep-passaged bovine spongiform encephalopathy agent. PLoS One 13:e0199914

Hermann P, Appleby B, Brandel J-P, Caughey B, Collins S, Geschwind MD, Green A, Haïk S, Kovacs GG, Ladogana A (2021) Biomarkers and diagnostic guidelines for sporadic Creutzfeldt-Jakob disease. Lancet Neurol 20:235–246

Hiller H, Yang C, Beachy DE, Kusmartseva I, Candelario-Jalil E, Posgai AL, Nick HS, Schatz D, Atkinson MA, Wasserfall CH (2021) Altered cellular localisation and expression, together with unconventional protein trafficking, of prion protein, PrP C, in type 1 diabetes. Diabetologia 64:2279–2291

Hu W, Kieseier B, Frohman E, Eagar TN, Rosenberg RN, Hartung H-P (2008) Prion proteins: physiological functions and role in neurological disorders. J Neurol Sci 264:1–8

Huarcaya SP, Zunke FJ (2023) Therapeutic potential of lysosomal cathepsins for neurodegenerative diseases. Neural Regen Res 18:1713

Hughes D, Halliday M (2017) What is our current understanding of PrP^{Sc}-associated neurotoxicity and its molecular underpinnings? Pathogens 6:63

Hutti CR, Welle KA, Hryhorenko JR, Ghaemmaghami SJ (2020) Global analysis of protein degradation in prion infected cells. Sci Rep 10:10800

Islam BU, Habib S, Ahmad P, Allarakha S, Ali A (2015) Pathophysiological role of peroxynitrite induced DNA damage in human diseases: a special focus on poly (ADP-ribose) polymerase (PARP). Indian J Clin Biochem 30:368–385

Jayanthi P, Thomas P, Bindhu P, Krishnapillai R (2013) Prion diseases in humans: oral and dental implications. N Am J Med Sci 5:399

Jellinger KA (2010) Basic mechanisms of neurodegeneration: a critical update. J Cell Mol Med 14:457–487

Ji CH, Kim HY, Heo AJ, Lee SH, Lee MJ, Kim SB, Srinivasrao G, Mun SR, Cha-Molstad H, Ciechanover A (2019) The N-degron pathway mediates ER-phagy. Mol Cell 75:1058–1072

Justo AFO, Suemoto CKJ (2022) The modulation of neuroinflammation by inducible nitric oxide synthase. Signaling 16:155–158

Kanemura S, Matsusaki M, Inaba K, Okumura MJ (2020) PDI family members as guides for client folding and assembly. Int J Mol Sci 21:9351

Kevei É, Pokrzywa W, Hoppe T (2017) Repair or destruction—an intimate liaison between ubiquitin ligases and molecular chaperones in proteostasis. FEBS Lett 591:2616–2635

Khan NA, Asim M, El-Menyar A, Biswas KH, Rizoli S, Al-Thani H (2022) The evolving role of extracellular vesicles (exosomes) as biomarkers in traumatic brain injury: clinical perspectives and therapeutic implications. Front Aging Neurosci 14:933434

Kim M-O, Takada LT, Wong K, Forner SA, Geschwind MD (2018) Genetic PrP prion diseases. Cold Spring Harb Perspect Biol 10:a033134

Kosorinova D, Belay G, Zakova D, Stelzer M, Mitrova EJP (2021) Genetic risk factors of creutzfeldt-jakob disease in the population of newborns in Slovakia. Pathogens 10:435

Kraus A, Raymond GJ, Race B, Campbell KJ, Hughson AG, Anson KJ, Raymond LD, Caughey BJ (2017) PrP P102L and nearby lysine mutations promote spontaneous in vitro formation of transmissible prions. J Virol 91:e01276

Kwon GT, Kwon MSJ (2019) Diagnostic challenge of rapidly progressing sporadic Creutzfeldt-Jakob disease. BMJ Case Rep 12:e230535

Lambert ZJ, Greenlee JJ, Cassmann ED, West Greenlee MH (2021) Differential accumulation of misfolded prion strains in natural hosts of prion diseases. Viruses 13:2453

Laskowska E, Kuczyńska-Wiśnik D, Lipińska BJ (2019) Proteomic analysis of protein homeostasis and aggregation. J Proteome 198:98–112

Lee Y-C (2023) Prions: a threat to health security and the need for effective medical countermeasures. Glob Health J 7:43–48

Lin Z, Zhao D, Yang L (2013) Interaction between misfolded PrP and the ubiquitin-proteasome system in prion-mediated neurodegeneration. Acta Biochim Biophys Sin 45:477–484

Liu Y, Tan L, Tan M-S (2023) Chaperone-mediated autophagy in neurodegenerative diseases: mechanisms and therapy. Biochemistry 11(14):2250

Llorens F, Thüne K, Sikorska B, Schmitz M, Tahir W, Fernández-Borges N, Cramm M, Gotzmann N, Carmona M, Streichenberger N (2017) Altered Ca^{2+} homeostasis induces calpain-cathepsin axis activation in sporadic Creutzfeldt-Jakob disease. Acta Neuropathol Commun 5:1–20

Lopes DM, Llewellyn SK, Harrison IF (2022) Propagation of tau and α-synuclein in the brain: therapeutic potential of the glymphatic system. Transl Neurodegener 11:1–23

López-Pérez Ó, Badiola JJ, Bolea R, Ferrer I, Llorens F, Martín-Burriel I (2020) An update on autophagy in prion diseases. Biotechnology 8:975

Lundberg JO, Weitzberg E (2022) Nitric oxide signaling in health and disease. Cell 185:2853–2878

Ma J, Gao J, Wang J, Xie AJ (2019) Prion-like mechanisms in Parkinson's disease. Front Neurosci 13:552

Majd S, Power JH, Grantham HJ (2015) Neuronal response in Alzheimer's and Parkinson's disease: the effect of toxic proteins on intracellular pathways. BMC Neurosci 16:1–13

Malgieri G, Grasso G (2014) The clearance of misfolded proteins in neurodegenerative diseases by zinc metalloproteases: an inorganic perspective. Coord Chem Rev 260:139–155

Marrero-Winkens C, Sankaran C, Schätzl HM (2020) From seeds to fibrils and back: fragmentation as an overlooked step in the propagation of prions and prion-like proteins. Biomolecules 10:1305

Maurya SK, Gupta S, Bakshi A, Kaur H, Jain A, Senapati S, Baghel MSJ (2022) Targeting mitochondria in the regulation of neurodegenerative diseases: a comprehensive review. J Neurosci Res 100:1845–1861

Michel V, Bakovic M (2007) Lipid rafts in health and disease. Biol Cell 99:129–140

Milhavet O, McMahon HE, Rachidi W, Nishida N, Katamine S, Mangé A, Arlotto M, Casanova D, Riondel J, Favier A (2000) Prion infection impairs the cellular response to oxidative stress. Proc Natl Acad Sci USA 97:13937–13942

Nakamura T, Oh C-K, Zhang X, Lipton SAJ (2021) Protein S-nitrosylation and oxidation contribute to protein misfolding in neurodegeneration. Medicine 172:562–577

Nakaoke R, Sakaguchi S, Atarashi R, Nishida N, Arima K, Shigematsu K, Katamine S (2000) Early appearance but lagged accumulation of detergent-insoluble prion protein in the brains of mice inoculated with a mouse-adapted Creutzfeldt–Jakob disease agent. Neurobiology 20:717–730

Olech MJ (2023) Conventional and state-of-the-art detection methods of bovine spongiform encephalopathy (BSE). Int J Mol Sci 24:7135

Orrù CD, Soldau K, Cordano C, Llibre-Guerra J, Green AJ, Sanchez H, Groveman BR, Edland SD, Safar JG, Lin JH (2018) Prion seeds distribute throughout the eyes of sporadic Creutzfeldt-Jakob disease patients. MBio 9:6

Otero A, Betancor M, Eraña H, Fernández Borges N, Lucas JJ, Badiola JJ, Castilla J, Bolea RJ (2021) Prion-associated neurodegeneration causes both endoplasmic reticulum stress and proteasome impairment in a murine model of spontaneous disease. Int J Mol Sci 22:465

Panda C, Mahapatra RK (2023) Bi-directional relationship between autophagy and inflammasomes in neurodegenerative disorders. Neurobiology 43:115–137

Panes JD, Saavedra P, Pineda B, Escobar K, Cuevas ME, Moraga-Cid G, Fuentealba J, Rivas CI, Rezaei H, Muñoz-Montesino CJ (2021) PrP C as a transducer of physiological and pathological signals. Front Mol Neurosci 14:762918

Peggion C, Stella R, Lorenzon P, Spisni E, Bertoli A, Massimino ML (2020) Microglia in prion diseases: angels or demons? Int J Mol Sci 21:7765

Peralta OA, Eyestone WH (2009) Quantitative and qualitative analysis of cellular prion protein (PrPC) expression in bovine somatic tissues. Prion 3:161–170

Pérez Gómez AA, Karmakar M, Carroll RJ, Lawley KS, Amstalden K, Young CR, Threadgill DW, Welsh CJ, Brinkmeyer-Langford C (2022) Serum cytokines predict neurological damage in genetically diverse mouse models. Cells 11:2044

Pisoschi AM, Pop A, Iordache F, Stanca L, Predoi G, Serban AI (2021) Oxidative stress mitigation by antioxidants-an overview on their chemistry and influences on health status. Eur J Med Chem 209:112891

Poggiolini I, Saverioni D, Parchi PJ (2013) Prion protein misfolding, strains, and neurotoxicity: an update from studies on mammalian prions. Int J Cell Biol 2013:910314

Polyakova O, Dear D, Stern I, Martin S, Hirst E, Bawumia S, Nash A, Dodson G, Bronstein I, Bayley PM (2009) Proteolysis of prion protein by cathepsin S generates a soluble β-structured intermediate oligomeric form, with potential implications for neurotoxic mechanisms. Eur Biophys J 38:209–218

Porter SR (2003) Prion disease: possible implications for oral health care. J Am Dent Assoc 134:1486–1491

Prasad KN, Bondy SC (2018) Oxidative and inflammatory events in prion diseases: can they be therapeutic targets? Curr Aging Sci 11:216–225

Rahman A, Saikia B, Gogoi CR, Baruah A (2022) Advances in the understanding of protein misfolding and aggregation through molecular dynamics simulation. Prog Biophys Mol Biol 175:31–48

Ransohoff RMJS (2016) How neuroinflammation contributes to neurodegeneration. Science 353:777–783

Rempe RG, Hartz AM, Bauer BJ (2016) Matrix metalloproteinases in the brain and blood–brain barrier: versatile breakers and makers. Metabolism 36:1481–1507

Sanz-Hernández M, Barritt JD, Sobek J, Hornemann S, Aguzzi A, De Simone A (2021) Mechanism of misfolding of the human prion protein revealed by a pathological mutation. Proc Natl Acad Sci USA 118:e2019631118

Sarnataro D, Pepe A, Zurzolo C (2017) Cell biology of prion protein. Science 150:57–82

Shaked GM, Fridlander G, Meiner Z, Taraboulos A, Gabizon RJ (1999) Protease-resistant and detergent-insoluble prion protein is not necessarily associated with prion infectivity. J Biol Chem 274:17981–17986

Shrestha P, Katila N, Lee S, Seo JH, Jeong J-H, Yook S (2022) Methamphetamine induced neurotoxic diseases, molecular mechanism, and current treatment strategies. Biomed Pharmacother 154:113591

Sies H, Jones DP (2020) Reactive oxygen species (ROS) as pleiotropic physiological signalling agents. Nat Rev Mol Cell Biol 21:363–383

Singh N, Chaudhary S, Ashok A, Lindner E (2020) Prions and prion diseases: insights from the eye. Exp Eye Res 199:108200

Sitammagari KK, Masood W (2018) Creutzfeldt Jakob disease. N Engl J Med 378(4):e7

Spagnolli G, Rigoli M, Orioli S, Sevillano AM, Faccioli P, Wille H, Biasini E, Requena JR (2019) Full atomistic model of prion structure and conversion. Int J Mol Sci 15:e1007864

Spiers JG, Chen H-JC, Steinert JR (2023) Redox mechanisms and their pathological role in prion diseases: the road to ruin. PLoS Pathog 19:e1011309

Sweeney P, Park H, Baumann M, Dunlop J, Frydman J, Kopito R, McCampbell A, Leblanc G, Venkateswaran A, Nurmi A (2017) Protein misfolding in neurodegenerative diseases: implications and strategies. Transl Neurodegener 6:1–13

Takada LT, Kim MO, Cleveland RW, Wong K, Forner SA, Gala II, Fong JC, Geschwind MD (2017) Genetic prion disease: experience of a rapidly progressive dementia center in the United States and a review of the literature. Am J Med Genet B Neuropsychiatr Genet 174:36–69

Taraboulos A, Jendroska K, Serban D, Yang S-L, DeArmond SJ, Prusiner SB (1992) Regional mapping of prion proteins in brain. Proc Natl Acad Sci USA 89:7620–7624

Thellung S, Corsaro A, Dellacasagrande I, Nizzari M, Zambito M, Florio TJ (2022) Proteostasis unbalance in prion diseases: mechanisms of neurodegeneration and therapeutic targets. Front Neurosci 16:966019

Thomzig A, Schulz-Schaeffer W, Wrede A, Wemheuer W, Brenig B, Kratzel C, Lemmer K, Beekes M (2007) Accumulation of pathological prion protein PrPSc in the skin of animals with experimental and natural scrapie. PLoS Pathog 3:e66

Thüne K, Schmitz M, Villar-Piqué A, Altmeppen HC, Schlomm M, Zafar S, Glatzel M, Llorens F, Zerr I (2019) The cellular prion protein and its derived fragments in human prion diseases and their role as potential biomarkers. Expert Rev Mol Diagn 19:1007–1018

Ugalde CL, Finkelstein DI, Lawson VA, Hill AFJ (2016) Pathogenic mechanisms of prion protein, amyloid-β and α-synuclein misfolding: the prion concept and neurotoxicity of protein oligomers. J Neurochem 139:162–180

Vafadari B, Salamian A, Kaczmarek LJ (2016) MMP-9 in translation: from molecule to brain physiology, pathology, and therapy. J Neurochem 139:91–114

Vanni S, Colini Baldeschi A, Zattoni M, Legname GJ (2020) Brain aging: a Ianus-faced player between health and neurodegeneration. J Neurosci Res 98:299–311

Villar-Piqué A, Schmitz M, Lachmann I, Karch A, Calero O, Stehmann C, Sarros S, Ladogana A, Poleggi A, Santana I (2019) Cerebrospinal fluid total prion protein in the spectrum of prion diseases. Mol Neurobiol 56:2811–2821

Wang Z, Manca M, Foutz A, Camacho MV, Raymond GJ, Race B, Orru CD, Yuan J, Shen P, Li B (2019) Early preclinical detection of prions in the skin of prion-infected animals. Nat Commun 10:247

Wang F, Yuan Q, Chen F, Pang J, Pan C, Xu F, Chen YJ (2021) Fundamental mechanisms of the cell death caused by nitrosative stress. Biology 9:742483

Watson N, Hermann P, Ladogana A, Denouel A, Baiardi S, Colaizzo E, Giaccone G, Glatzel M, Green AJ, Haïk SJ (2022) Validation of revised international Creutzfeldt-Jakob disease surveillance network diagnostic criteria for sporadic Creutzfeldt-Jakob disease. JAMA Netw Open 5:e2146319

Watts JC, Bourkas ME (2018) The function of the cellular prion protein in health and disease. Acta Neuropathol 135:159–178

Watts JC, Huo H, Bai Y, Ehsani S, Won AH, Shi T, Daude N, Lau A, Young R, Xu L (2009) Interactome analyses identify ties of prpc and its mammalian paralogs to oligomannosidic n-glycans and endoplasmic reticulum-derived chaperones. PLoS Pathog 5:e1000608

Weaver K, Skouta RJB (2022) The selenoprotein glutathione peroxidase 4: from molecular mechanisms to novel therapeutic opportunities. Biomedicine 10:891

Weissmann C, Enari M, Klöhn P-C, Rossi D, Flechsig EJ (2002) Transmission of prions. J Infect Dis 186:S157–S165

Wentink A, Nussbaum-Krammer C, Bukau B (2019) Modulation of amyloid states by molecular chaperones. Cold Spring Harb Perspect Biol 11:a033969

Westergard L, Christensen HM, Harris DA (2007) The cellular prion protein (PrPC): its physiological function and role in disease. Biochim Biophys Acta 1772:629–644

Wille H, Requena JR (2018) The structure of PrPSc prions. Pathogens 7:20
Xu Y, Xu J, Zhang J, Cai Z, Wei H, Yu M, Li Y (2019) Sporadic Creutzfeldt–Jakob disease presenting as dizziness and cognitive decline: a case report. Medicine 98:e16002
Yadav D, Lee JY, Puranik N, Chauhan PS, Chavda V, Jin J-O, Lee PC (2022) Modulating the ubiquitin–proteasome system: a therapeutic strategy for autoimmune diseases. Cells 11:1093
Yadavalli R, Guttmann RP, Seward T, Centers AP, Williamson RA, Telling GCJ (2004) Calpain-dependent endoproteolytic cleavage of PrPSc modulates scrapie prion propagation. J Biol Chem 279:21948–21956
Yagami T, Yamamoto Y, Koma H (2019) Pathophysiological roles of intracellular proteases in neuronal development and neurological diseases. Mol Neurobiol 56:3090–3112
Yoon S, Eom GH, Kang GJ (2021) Nitrosative stress and human disease: therapeutic potential of denitrosylation. Int J Mol Sci 22:9794
Yoshida S, Hasegawa T (2022) Deciphering the prion-like behavior of pathogenic protein aggregates in neurodegenerative diseases. Neurochem Int 155:105307
Yuan AH, Garrity SJ, Nako E, Hochschild AJE (2014) Prion propagation can occur in a prokaryote and requires the ClpB chaperone. elife 3:e02949
Yusuf M, Khan M, Robaian MA, Khan RA (2018) Biomechanistic insights into the roles of oxidative stress in generating complex neurological disorders. Biol Chem 399:305–319
Zakany F, Mándity IM, Varga Z, Panyi G, Nagy P, Kovacs T (2023) Effect of the lipid landscape on the efficacy of cell-penetrating peptides. Cells 12:1700
Zeng L, Zou W, Wang GJ (2015) Cellular prion protein (PrPC) and its role in stress responses. Int J Clin Exp Med 8:8042
Zhang W, Xiao D, Mao Q, Xia HJST, Therapy T (2023) Role of neuroinflammation in neurodegeneration development. Signal Transduct Target Ther 8:267
Zhu T, Chen J-L, Wang Q, Shao W, Qi B (2018) Modulation of mitochondrial dynamics in neurodegenerative diseases: an insight into prion diseases. Front Aging Neurosci 10:336
Zou W-Q (2023) Insoluble cellular prion protein and other neurodegeneration-related protein aggregates in the brain of asymptomatic individuals. In: Prions and diseases. Springer, Cham, pp 63–86

Modulation of Several Downstream Cascades Served by Enzymes in the Pathogenesis of Stroke

9

Sayed Mohammed Firdous and Sourav Pal

Abstract

Interruption of blood circulation to the brain, leading to neuronal destruction and functional impairment, is an attribute of a stroke, a major contributor to death and disability worldwide. The enzymes associated with oxidative assault, inflammation, and tissue repair are all crucial in understanding the pathophysiology of stroke. Oxygen deprivation during a stroke causes enzyme activation, which in turn causes the production of reactive oxygen species (ROS) and oxidative stress. ROS, generated by nicotinamide adenine dinucleotide phosphate and xanthine oxidases, exacerbates brain cell injury by inducing lipid peroxidation, DNA damage, and protein oxidation. Inflammatory mediators are produced when cyclooxygenase (COX), lipoxygenase (LOX), and matrix metalloproteinases (MMPs) are activated. COX enzymes generate prostaglandins, which promote inflammation and contribute to brain injury. LOX enzymes produce leukotrienes and other lipid mediators that exacerbate inflammation and tissue damage. MMPs, particularly MMP-9, contribute to blood-brain barrier disruption and inflammatory cell infiltration. Excitotoxicity, the pathological process involving excessive stimulation of glutamate receptors, contributes to neuronal death in stroke. Excitotoxic signaling is accomplished by glutamate receptors like N-methyl-D-aspartate (NMDA) and AMPA receptors, and glutamate-activated enzymes and GSs regulate glutamate metabolism. Tissue remodeling and removal of cellular waste aid tPA and MMPs. Patients experiencing an ischemic stroke are given thrombolytic therapy with tPA. In addition, MMP activation pro-

S. M. Firdous (✉)
Department of Pharmacology, Calcutta Institute of Pharmaceutical Technology & AHS, Howrah, West Bengal, India

S. Pal
Department of Pharmacology, Seacom Pharmacy College, Jaladhulagori, Sankrail, West Bengal, India

B. Mathew, D. G. T. Parambi (eds.), *Enzymes in Neurodegenerative Disorders*,
https://doi.org/10.1007/978-981-97-6822-6_9

motes extracellular matrix (ECM) disintegration, which aids in tissue repair and functional restoration. Enzyme research into stroke has provided new targets for treatment. In cases of ischemic stroke, thrombolytic therapy like tPA injection tries to improve blood flow. Stroke patients' brain damage can be reduced and their outcomes improved by blocking inflammatory enzymes, including COX and MMPs. Finally, enzymes are involved in many different aspects of stroke, including oxidative assault, inflammation, and tissue healing. Understanding enzymes' intricate role in stroke pathogenesis provides opportunities to discern avenues for intervention that enhance patient outcomes.

Keywords

Stroke · NADPH oxidase · Xanthine oxidase · COX · LOX · MMPs · Drp1

9.1 Introduction

When the blood supply to the brain is abruptly interrupted, brain cells perish, and neurological damage occurs; this situation is known as a stroke. Millions of individuals around the world are affected by stroke (Chemerinski and Robinson 2000). The World Health Organization estimates 15 million people worldwide suffer a stroke annually, with 5 million dying as a direct result. Regarding global prevalence, stroke ranks at number two for mortality and number three for disability. Stroke rates in India range considerably between geographical areas and demographic categories. While exact numbers are not available, research places the annual incidence rate anywhere from 119 to 145 per 100,000 people (Das et al. 2007). The incidence tends to be higher in urban areas than in rural areas, likely due to lifestyle factors and access to healthcare. While the primary cause of stroke is the blockage or rupture of blood vessels supplying the brain, the resulting ischemia (lack of oxygen and nutrients) triggers a complex series of pathological events known as the downstream cascade. The induction of oxidative assault, which is characterized by an imbalance between the creation of ROS and the potential of the body's antioxidant defenses to neutralize them, is a critical aspect of the downstream cascade in stroke (Chavda et al. 2022). Several key enzymes play a significant role in this pathological process, including nicotinamide adenine dinucleotide phosphate (NADPH) oxidase, xanthine oxidase (XO), cyclooxygenase (COX), lipoxygenase (LOX), matrix metalloproteinase-9 (MMP-9), glutathione decarboxylase, and glutathione synthase (Mohamed et al. 2019). NADPH oxidase is an enzyme complex found in various cell types, including neurons, endothelial cells, and immune cells. Activation of NADPH oxidase in response to ischemia injury results in the generation of superoxide radicals ($O_2^{\bullet -}$) and other ROS. In turn, these ROS lead to oxidative degradation of lipids, proteins, and DNA within cells. Another enzyme associated with stroke-induced oxidative assault is XO. ROS are produced in the process of transforming hypoxanthine into xanthine and then uric acid. Ischemia causes a change in purine metabolism, with more XO activity and, consequently, more reactive oxygen species (ROS) generation and oxidative stress. Arachidonic acid is metabolized by the enzymes COX and LOX, both of which are found in cell membranes (Hotter et al. 1995). After a stroke, the triggering of these enzymes results in the synthesis of prostaglandins and leukotrienes, two types of inflammatory mediators. MMP-9 is an enzyme involved in the breakdown of the ECM, which provides structural support to brain tissue (Kelly et al. 2018). In stroke, MMP-9 is upregulated, leading to increased degradation of the ECM components. This process disrupts the blood-brain barrier (BBB), facilitates inflammation, and contributes to neuronal death. Glutathione decarboxylase and glutathione synthase are enzymes involved in synthesizing and recycling glutathione, a critical antioxidant molecule in the body. Stroke disrupts the redox balance and depletes glutathione levels, impairing the antioxidant defenses. This depletion further exacerbates oxidative stress and neuronal death (Higashi et al. 2021).

9.2 NADPH Oxidase

Ischemia-reperfusion injury is the damage done to tissues due to an intermittent halt (ischemia) and subsequent restoration (reperfusion) of blood circulation. The molecular basis of ischemia-reperfusion injury involves a complex cascade of events contributing to tissue damage and dysfunction. The molecular cascade of NADPH oxidase activation in ischemia-reperfusion injury involves a series of events that contribute to oxidative stress and tissue damage (Granger and Kvietys 2015). Here is an overview of this cascade:

During the ischemic phase of injury, blood flow to a specific tissue or organ is disrupted, leading to oxygen and nutrient deprivation. The lack of oxygen triggers a complex series of molecular and cellular responses. Ischemia activates NADPH oxidase, primarily in immune and endothelial cells in the affected tissue. The activation process involves the assembly of a multi-subunit enzyme complex. Substantially shifting electrons from NADPH to molecular oxygen, the activated NADPH oxidase produces superoxide radicals ($O_2^{\bullet-}$) (Loukogeorgakis et al. 2010). The electron transfer is facilitated by a complex enzyme composed of membrane-bound subunits (like gp91phox and p22phox) and cytosolic subunits (like p47phox, p67phox, and p40phox) (Zeng et al. 2019). Disputation of the produced superoxide radicals into other ROS, most notably hydrogen peroxide (H_2O_2), is possible through spontaneous or enzymatic processes. Superoxide dismutase (SOD) is an enzyme that helps with the transformation. Reperfusion occurs when blood flow is restored to the ischemic tissue. Paradoxically, reperfusion can exacerbate tissue injury due to the generation of additional ROS and oxidative assault. The increased production of ROS, including superoxide and H_2O_2, contributes to oxidative stress and cellular damage (Fig. 9.1). ROS can directly damage lipids, proteins, and DNA, leading to oxidative modifications and dysfunction of cellular components. ROS generated by NADPH oxidase activation triggers the activation of various signaling pathways involved in inflammation, such as nuclear factor-kappa B (NF-κB) and mitogen-activated protein kinases (MAPKs). This releases pro-inflammatory cytokines, chemokines, and adhesion molecules, amplifying the inflammatory response and tissue damage.

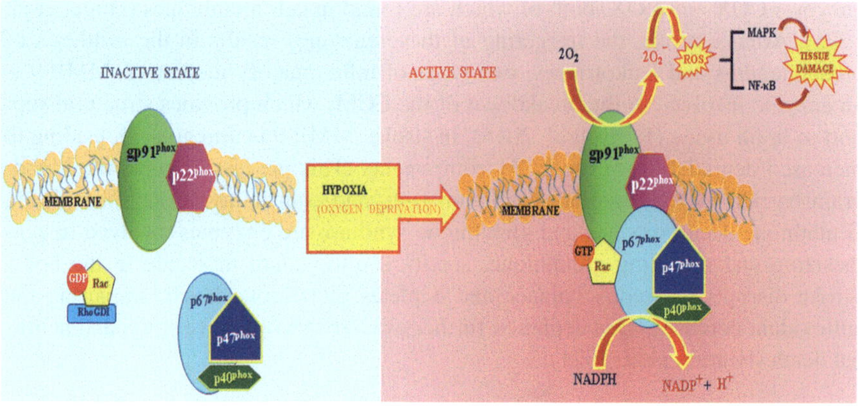

Fig. 9.1 Schematic depiction of the inactive and active forms of NADPH oxidase

Oxidative assault and ROS production during ischemia-reperfusion injury can impair mitochondrial function. Mitochondrial dysfunction contributes to the further generation of ROS, disruption of energy metabolism, and release of pro-apoptotic factors, leading to cell death. The combination of oxidative stress, inflammation, and mitochondrial dysfunction can trigger various cell death pathways, including apoptosis, necrosis, and autophagy, depending on the severity and duration of the injury (Matsushima et al. 2014).

The pathophysiology of brain stroke and neuronal death is intricately entwined with the molecular downstream cascade of NADPH oxidase-induced oxidative and nitrosative stress. This chain reaction can be summarized as follows:

- NADPH oxidase must first be converted from its inactive form as a membrane-bound complex to be active. A membrane-bound flavocytochrome b558, composed of gp91phox (also known as Nox2) and p22phox, is a crucial component of the complex, as are many cytosolic subunits, including as p47phox, p67phox, and a small GTPase called Rac. Neurons, endothelial cells, and immune cells all activate NADPH oxidase in response to pathogenic stimuli following brain stroke, such as ischemia-reperfusion injury. This activation process begins with forming a multi-subunit enzyme complex and ends with the migration of electrons from NADPH to atomic oxygen. Upon receiving activation signals, cytosolic subunits of flavocytochrome b558 are shuttled to the membrane-bound form of the protein. The phosphorylation of p47phox in the cytosol causes conformational changes that facilitate its liaison with flavocytochrome b558. When these two molecules connect, it helps bring together the activated NADPH oxidase complex (Filip-Ciubotaru et al. 2016).
- ROS Generation: Upon activation, NADPH oxidase produces superoxide radicals ($O_2^{\bullet-}$) as its major product, resulting in ROS. The electron transfers from NADPH to oxygen results in the formation of $O_2^{\bullet-}$. This superoxide production occurs in the cellular compartments where NADPH oxidase is localized, such as

the plasma membrane or endosomes. The produced superoxide radicals can undergo dismutation either spontaneously or through enzymatic reactions, resulting in the generation of other ROS, particularly H_2O_2. This conversion is facilitated by the enzyme SOD. H_2O_2 serves as a stable ROS and can further contribute to oxidative stress and neuronal damage (Parascandolo and Laukkanen 2019).

- Nitric oxide (NO) interaction: NADPH oxidase-derived superoxide radicals can react with NO expressed by neurons and endothelial cells. This reaction leads to the formation of peroxynitrite (ONOO–), a highly reactive species with strong oxidizing and nitrating capabilities. Peroxynitrite contributes to nitrosative stress and further amplifies oxidative damage in stroke.
- Oxidative and nitrosative stress: The increased production of ROS, particularly H_2O_2, and the formation of ONOO– contribute to oxidative and nitrosative stress in the brain. These reactive species can directly damage lipids, proteins, and DNA within neurons, leading to oxidative modifications and cellular dysfunction. Oxidative and nitrosative stress disrupts cellular homeostasis and contributes to the progression of brain injury.
- Activation of inflammatory pathways: NADPH oxidase-derived ROS and peroxynitrite activate various inflammatory pathways, such as NF-κB and MAPKs. These cascades trigger the release of pro-inflammatory cytokines, chemokines, and adhesion molecules, promoting neuroinflammation and exacerbating brain injury in stroke (Radi 2018).
- Mitochondrial dysfunction and apoptosis: Oxidative and nitrosative stress resulting from NADPH oxidase activation leads to mitochondrial dysfunction, including impaired mitochondrial respiration, increased production of ROS within mitochondria, and mitochondrial membrane damage. These events disrupt energy production and trigger apoptotic signaling pathways, ultimately leading to neuronal death in stroke (Tian et al. 2022).

During reperfusion, the sudden increase in oxygen availability can lead to the generation of excessive ROS. Inflammation, mitochondrial malfunction, and cell death can all result from the oxidation of lipids, proteins, and DNA by these ROS. Blood clots can be dismantled with the help of an enzyme called tissue plasminogen activator (tPA). In the case of an ischemic stroke, it is used effectively to dissolve the clot blocking a blood vessel and restore blood flow. However, tPA administration during reperfusion can potentially contribute to oxidative stress and tissue damage. Studies have suggested that tPA can increase the production of ROS during reperfusion. One mechanism proposed for this effect is the activation of NADPH oxidase, an enzyme complex that generates superoxide radicals. tPA can activate NADPH oxidase in various cell types, including endothelial cells and neurons, leading to increased ROS generation and oxidative assault. The excess ROS generated during hypoxia-reperfusion injury, whether from tPA or other sources, can further contribute to tissue damage. ROS can initiate lipid peroxidation, protein oxidation, and DNA damage, impair antioxidant defense systems, and promote inflammation. This oxidative stress can exacerbate cell injury and contribute to the pathogenesis of conditions such as ischemic stroke (Hermann and Matter 2007).

9.3 XO

The activation of XO can occur through various mechanisms, including hypoxia, inflammation, ischemia-reperfusion injury, and increased levels of purine substrates (Berry and Hare 2004). Here's a description of the molecular cascade involved in the activation of XO by these factors:

In the absence of oxygen, or hypoxia, hypoxia-inducible factor (HIF) becomes stabilized. In the presence of oxygen, HIF is hydroxylated and subsequently degraded. Because hydroxylation is blocked in hypoxia, HIF levels rise. In the nucleus, stabilized HIF triggers the transcription of several genes, including XO. The XO gene is activated when HIF attaches to hypoxia response elements in its promoter. During an inflammatory reaction, cytokines such as tumor necrosis factor-alpha and interleukin (IL)-1 are released. The transcription factor NF-B is triggered by inflammatory cytokines. Following this event, NF-kB moves into the nucleus and attaches to the promoter region of the XO gene, increasing the production of that molecule. Inflammation-induced ROS production can directly activate XO or promote its conversion from the inactive form (xanthine dehydrogenase) to the active form (XO). ROS generated by other sources can also contribute to XO activation (Takeda et al. 2010).

During ischemia, blood flow to a specific tissue or organ is interrupted, leading to oxygen and nutrient deprivation. Restoration of blood flow (reperfusion) following ischemia results in a sudden increase in oxygen availability. The sudden increase in oxygen availability during reperfusion can lead to the activation of XO. The influx of oxygen facilitates the conversion of xanthine dehydrogenase to XO, which generates superoxide radicals ($O_2^{\bullet-}$) and other reactive species.

Increased purine substrates, such as hypoxanthine and xanthine, can occur in conditions like tissue damage, trauma, or certain metabolic disorders. Elevated levels of purine substrates can stimulate XO activity. By acting as a catalyst, XO converts hypoxanthine to xanthine and then uric acid, with the byproduct of superoxide radicals.

Here's an overview of the molecular downstream cascade of –XO-induced oxidative and nitrosative assault in the pathogenesis of brain stroke and neuronal death:

XO activation leads to the generation of ROS as byproducts of its enzymatic activity. These ROS, particularly superoxide radicals ($O_2^{\bullet-}$), can directly target mitochondria and disrupt their function. Excessive ROS production can overwhelm the mitochondrial antioxidant defense systems, leading to lipid peroxidation, membrane destabilization, and permeabilization. Oxidative assault on mitochondrial proteins can impair their structure and function, affecting critical processes such as electron transport chain (ETC) activity and adenosine triphosphate (ATP) production (Shirley et al. 2014). Oxidative assault on mitochondrial DNA can lead to mutations and deletions, compromising mitochondrial function. This disruption impairs the flow of electrons, reducing ATP production and leading to an imbalance in the redox status of mitochondria. XO activation can also affect calcium homeostasis within the mitochondria. Calcium overload, induced by ROS, can impair mitochondrial calcium handling mechanisms and disrupt normal calcium signaling.

Elevated mitochondrial calcium levels can lead to mitochondrial dysfunction, affecting ATP production and triggering cell death pathways. Excessive or prolonged mitophagy activation can lead to the depletion of functional mitochondria and further compromise cellular energy production and subsequently neuronal death (Guo et al. 2013).

9.4 COX

COX enzymes play a significant role in triggering the expression of prostaglandins, which are lipid mediators involved in various physiological and pathological processes, including neuro-inflammation and brain injury. Here is an overview of the molecular cascade: In response to inflammatory signals, such as cytokines, growth factors, or tissue damage, COX enzymes, specifically COX-1 and COX-2, are activated. COX-2 is primarily induced in response to inflammatory stimuli (Hickey et al. 2007). COX enzymes convert arachidonic acid, a polyunsaturated fatty acid present in the cell membrane, into prostaglandin H2 (PGH2). Arachidonic acid is released from membrane phospholipids through phospholipase A2 (PLA2) action. PGH2, generated by COX enzymes, serves as a precursor for various prostaglandins, such as prostaglandin E2 (PGE2), prostaglandin D2 (PGD2), and prostaglandin F2-alpha (PGF2α). These prostaglandins exert diverse effects on inflammation, vasodilation, immune responses, and neuronal signaling. Prostaglandins, particularly PGE2, have potent pro-inflammatory effects, leading to the production and release of inflammatory mediators (Caron et al. 2016). Immune cells like macrophages and neutrophils are recruited and activated partly because they upregulate adhesion molecule expression, which facilitates immune cell adherence and migration across the BBB and into the brain. Prostaglandins released during neuro-inflammation act on various cell types, including neurons, glial cells (astrocytes, microglia), and endothelial cells. They promote the production and release of pro-inflammatory cytokines, such as tumor necrosis factor-alpha (TNF-α), interleukin-1 beta (IL-1β), and IL-6. These cytokines further amplify the inflammatory response and contribute to neuronal damage. Prostaglandins can affect the integrity of the BBB, which generally restricts the entry of immune cells and molecules into the brain (Ghazanfari et al. 2021). Prostaglandins promote the expression of MMPs and other factors that can degrade the tight junction proteins, leading to BBB disruption and increased permeability. Neuroinflammation and BBB disruption facilitated by prostaglandins contribute to neuronal injury. The inflammatory mediators and immune cells released during neuro-inflammation can damage neurons and disrupt neuronal function. Additionally, the breakdown of the BBB allows the infiltration of cytotoxic substances and immune cells into the brain, further exacerbating neuronal damage (Engblom et al. 2002).

9.5 LOX

These enzymes trigger the expression of leukotrienes, which can contribute to neuro-inflammation and brain injury. Here is an overview of the molecular cascade:

LOX enzymes, particularly 5-lipoxygenase (5-LOX), are activated in response to various inflammatory signals and stimuli. It is the rate-limiting enzyme in leukotriene biosynthesis encoded by the ALOX gene. LOX enzymes use arachidonic acid, a polyunsaturated fatty acid, as a substrate. Lipases like phospholipase A2 (PLA2) contribute to its release from membrane phospholipids. LOX enzymes, including 5-LOX, metabolize arachidonic acid and convert it into leukotriene A4 (LTA4), the initial product of the leukotriene synthesis pathway. Specific enzymes can further metabolize LTA4 to generate different types of leukotrienes, including leukotriene B4 (LTB4) and cysteinyl leukotrienes (LTC4, LTD4, and LTE4). Leukotrienes, particularly LTB4 and cysteinyl leukotrienes have potent pro-inflammatory effects. LTB4 acts as a chemoattractant, promoting the recruitment and activation of immune cells, such as neutrophils, monocytes, and eosinophils, to the site of inflammation (Rådmark et al. 2015). Cysteinyl leukotrienes induce smooth muscle contraction, increase vascular permeability, and contribute to bronchoconstriction and other inflammatory responses (Chung 1995). The release of leukotrienes during neuro-inflammation can activate and attract immune cells, including microglia, astrocytes, and infiltrating leukocytes, to the site of brain injury or inflammation. Activated immune cells release pro-inflammatory mediators, such as cytokines (e.g., TNF-α, IL-1β) and ROS, further propagating the inflammatory response and contributing to neuronal damage (Wang et al. 2020). Leukotrienes can also impact the integrity of the BBB, similar to other inflammatory mediators. They can promote the expression of adhesion molecules and other factors contributing to BBB disruption, allowing immune cells and inflammatory mediators to penetrate the brain parenchyma more easily. Neuro-inflammation facilitated by leukotrienes, along with BBB disruption and the release of inflammatory mediators, can lead to neuronal injury. Immune cells and inflammatory mediators can directly damage neurons, initiate excitotoxicity, and induce oxidative stress, contributing to neuronal dysfunction and cell death (Wang et al. 2006).

9.6 MMP-9

Different cell types, such as endothelial cells, fibroblasts, and immune cells, synthesize the pro-enzyme form of MMP-9. The inactive form is secreted into the extracellular space. Pro-MMP-9 contains a pro-domain that acts as a regulatory domain, keeping the enzyme inactive (Michaluk and Kaczmarek 2007). The pro-domain prevents premature activation and indiscriminate proteolysis. Inflammatory signals, such as cytokines (e.g., tumor necrosis factor-alpha, IL-1) or oxidative stress, can trigger the release of these signals in the microenvironment (Malemud 2017). Activation signals lead to the activation of proteases, such as plasmin, serine proteases, or other MMPs, in the vicinity of pro-MMP-9. These proteases specifically recognize and cleave the pro-domain of pro-MMP-9 at specific sites. The proteolytic cleavage of the pro-domain results in removing the inhibitory region, enabling the active form of MMP-9 (Malemud 2017; Szklarczyk et al. 2002).

- Once activated and released, MMP-9 can degrade various components of the ECM near the BBB. It can specifically target and degrade ingredients like type IV collagen and laminin, which are essential for maintaining the integrity of the BBB (Malemud 2017; Szklarczyk et al. 2002; Ramos-DeSimone et al. 1999).
- Tight Junction Disruption: Tight associations between the endothelial cells that make up the BBB can also be broken down by MMP-9. By creating a rigid barrier between endothelial cells and obstructing the paracellular transport of molecules, tight junctions serve a critical function in controlling the permeability of the BBB. MMP-9 can degrade tight junction proteins, such as occludins and claudins, leading to the loosening of tight junction complexes and increased BBB permeability. In addition to targeting the tight junctions, MMP-9 can remodel the basement membrane, a specialized ECM layer underlying endothelial cells. MMP-9 can degrade components of the basement membrane, including laminin and type IV collagen, further compromising the structural integrity of the BBB (Brilha et al. 2017).
- Leukocyte Recruitment and Inflammation: MMP-9 can facilitate the recruitment and infiltration of leukocytes, mainly neutrophils, and monocytes, across the disrupted BBB. This recruitment is mediated by the generation of chemotactic signals and the degradation of ECM components. Inflammatory cells can exacerbate the inflammatory response in the brain, leading to tissue damage and injury (Kluger et al. 2013; Khandoga et al. 2006).
- Increased Vascular Permeability: The disruption of tight junctions and basement membrane remodeling caused by MMP-9 results in increased vascular permeability at the BBB. The brain parenchyma becomes permeable to immune cells, inflammatory mediators, and other potentially toxic substances, resulting in neuroinflammation and even more tissue damage. When the BBB is compromised, inflammatory cells and mediators can enter the brain and set off a chain reaction of inflammatory reactions, also known as neuroinflammation. Neuronal injury, oxidative stress, excitotoxicity, and ultimately brain damage are all aided by the release of pro-inflammatory cytokines, ROS, and other neurotoxic chemicals by activated immune cells (Fig. 9.2) (Turner and Sharp 2016).

9.7 Excitotoxicity

It is a pathogenic state resulting from the overstimulation of glutamate receptors on neurons due to an accumulation of glutamate in the synaptic cleft. Damage to neurons and other brain tissue is the end outcome of a chain reaction at the molecular level. Glutamate is essential to healthy brain function as the principal excitatory neurotransmitter in the CNS. However, when glutamate is released excessively or

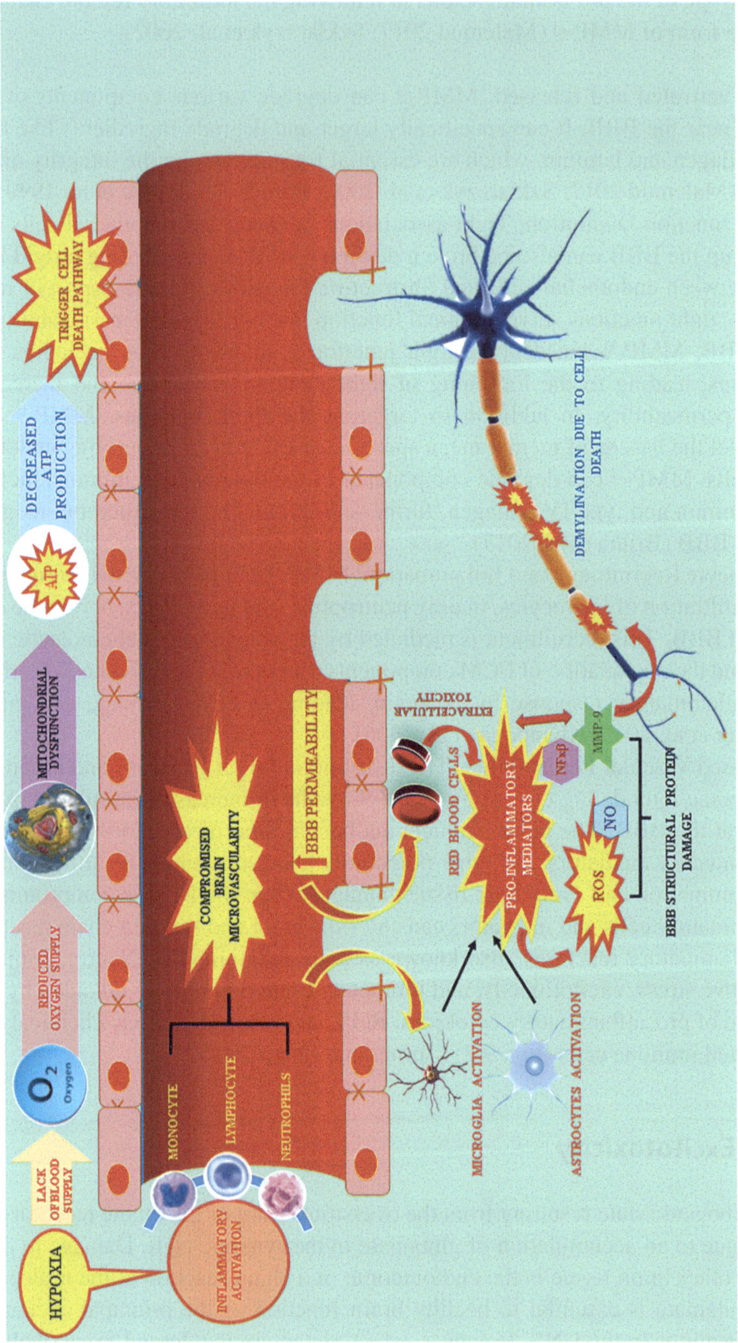

Fig. 9.2 MMP-9 mediated inflammatory activation and disruption of the blood-brain barrier

fails to be cleared from the synaptic cleft effectively, it can lead to excitotoxicity (Mark et al. 2001; Dong et al. 2009).

In response to neural activity, presynaptic neurons release glutamate into the synaptic cleft. Excessive stimulation of glutamate receptors, particularly NMDA receptors, occurs when there is an abundance of glutamate in the synaptic cleft. Ca^{2+} ions enter the postsynaptic neuron after NMDA receptor activation. When calcium ions flood a cell in excess, it can disrupt normal cellular processes and create a cascade of destructive events inside the neuron. Calcium excess causes mitochondrial malfunction, reducing ATP production and increasing ROS production. Damage to lipids, proteins, and DNA, among other biological components, results from oxidative stress brought on by increased ROS. Oxidative assault and calcium overload activate various proteases and lipases, leading to the degradation of cellular proteins and membranes. Pro-inflammatory molecules are discharged, glial cells (microglia and astrocytes) are triggered, and immune cells are attracted when excitotoxicity is prevalent. Damage to neurons is exacerbated by neuroinflammation (Tilleux and Hermans 2007; Candee et al. 2023).

9.8 Excitotoxicity and the Enzymes Glutamate-Activated Decarboxylase (GAD) and Glutamate Synthase

Glutamate metabolism involves enzymes such as GAD and glutamate synthetase. However, these enzymes are primarily interested in the synthesis and metabolism of glutamate rather than excitotoxicity (Dong et al. 2009). Gamma-aminobutyric acid (GABA) is the primary inhibitory neurotransmitter in the brain, and GAD is responsible for converting glutamate to GABA. Disturbance of the equilibrium between excitable and suppressive neurotransmission might result from GAD malfunction. GAD can be found in two different isoforms, namely, GAD65 and GAD67. These isoforms are named based on their approximately 65 and 67 kilodaltons molecular weights, respectively (Tavazzani et al. 2014; Popp et al. 2009). Both isoforms catalyze the decarboxylation of glutamate to GABA using the coenzyme pyridoxal phosphate. GAD is primarily found in GABAergic neurons, inhibitory neurons that utilize GABA as their primary neurotransmitter. GABAergic neurons provide inhibitory signals to other neurons by releasing GABA into the synapse. GABA then binds to specific GABA receptors on the postsynaptic neurons, decreasing their excitability. Dysfunction of GAD and subsequent disruption of the balance between excitatory and inhibitory neurotransmission can have significant implications for brain function (Nuss 2015). During a stroke, there is a disruption in blood supply to a specific area of the brain, either due to a blockage or hemorrhagic stroke. The damaged region suffers from a deficiency of oxygen and nutrients, which ultimately results in the degeneration of brain cells. GABA and GABA receptors are found in cerebral blood vessels, and GABAergic signaling has been shown to regulate cerebral blood flow. Changes in GAD activity or GABAergic signaling could impact blood vessel tone and cerebral blood flow, which play a crucial role in stroke pathophysiology. GABA has been shown to have neuroprotective effects in various experimental models. It can reduce oxidative stress, inhibit inflammation, and modulate calcium homeostasis, which is implicated in stroke-induced neuronal damage. GAD isoforms may contribute to these neuroprotective effects through GABA synthesis.

Glutamate synthetase, also known as glutamine synthetase, plays a role in the glutamate synthesis from glutamine. Glutamine is taken up by astrocytes, where it is converted to glutamate by glutamate synthetase. Glutamate can then be released by astrocytes or converted back to glutamine to maintain glutamate homeostasis. While abnormalities in glutamate metabolism can contribute to neurological disorders, they do not directly induce excitotoxicity or neuronal death. Excitotoxicity primarily occurs due to excessive glutamate release and impaired clearance, leading to the molecular cascade described earlier (Suarez et al. 2002).

9.9 Elevated Expression of Dynamin-Related Protein 1 (Drp1) After Stroke-Induced Cell Death

Drp1 is a protein involved in mitochondrial fission, essential for maintaining mitochondrial health and function. After a stroke, there is evidence that Drp1 expression increases, and this upregulation has been associated with neuronal injury and neurodegeneration (Li et al. 2022). In this process, engagement of the p38 mitogen-activated protein kinase (p38 MAPK) pathway has been shown to serve an aspect. Here is an overview of the molecular mechanism of increased Drp1 expression after stroke involving p38 and MAPK pathway activation (Li et al. 2022; Pegadraju et al. 2022):

A stroke leads to a disruption of blood flow to a brain region, resulting in oxygen and nutrient deprivation, leading to ischemic insult. Various signaling pathways, including the p38 MAPK pathway, are triggered during ischemic conditions. The p38 MAPK pathway is activated in response to cellular stress and inflammation. Ischemic insult activates p38 MAPK through phosphorylation, leading to its activation. Activated p38 MAPK can enter the nucleus and phosphorylate various downstream targets, including ATF2 and cAMP-responsive element-binding protein (CREB). These phosphorylated transcription factors bind to specific DNA sequences in the promoter region of the Drp1 gene. The phosphorylated ATF2 and CREB binding to the Drp1 gene promoter enhances the transcriptional activity of Drp1. This results in increased transcription and synthesis of Drp1 mRNA. Drp1 mRNA is translated in the cytoplasm, leading to the synthesis of increased levels of Drp1 protein. Elevated levels of Drp1 protein promote mitochondrial fission by recruiting to the outer mitochondrial membrane and forming spiral-like structures around the mitochondria. The discrepancy between mitochondrial fusion and fission causes mitochondria to become fragmented and mitochondrial function to be slashed. Excessive mitochondrial fission induced by increased Drp1 expression contributes to neuronal injury and neurodegeneration after stroke (Fig. 9.3). Fragmented mitochondria impair bioenergetics, increase ROS production, and release pro-apoptotic

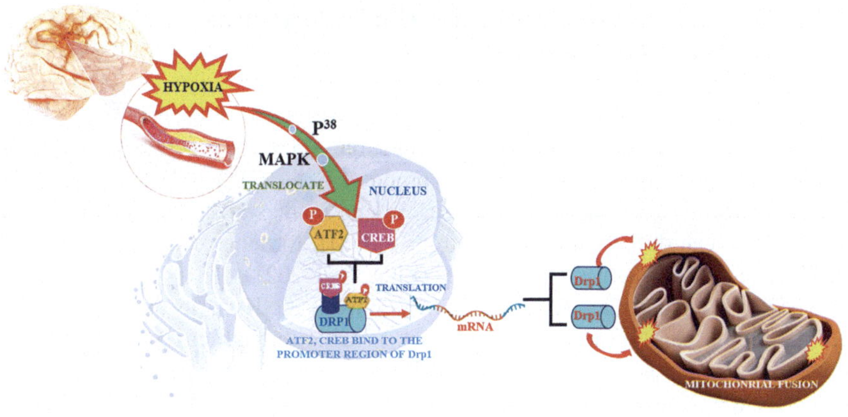

Fig. 9.3 Drp1 induced mitochondrial fission

factors, leading to cell death and exacerbation of brain damage (Pegadraju et al. 2022; Vongsfak et al. 2021).

9.10 Elevated Drp1 Expression-Induced Enzymatic Activation and Apoptosis

Elevated levels of Drp1 protein result in increased mitochondrial fission, leading to the fragmentation of mitochondria into smaller units. This excessive fission disrupts the balance between mitochondrial fusion and fission processes. Fragmented mitochondria exhibit impaired bioenergetics and reduced ATP production. Additionally, the fragmented mitochondrial network is less efficient in buffering calcium ions, leading to calcium overload in the mitochondrial matrix. Dysfunctional mitochondria produce increased levels of ROS as a byproduct of ETC activity. Oxidative assault on cells can be caused by ROS, such as $O_2^{\bullet-}$, $OH^{\bullet}$. In addition to calcium excess, ROS can cause the outer mitochondrial membrane permeability transition pore (mPTP) to open. Consequently, pro-apoptotic substances are released from the intermembrane gap into the cytosol, and the mitochondrial membrane potential is lost. The apoptosome is composed of cytochrome c, apoptotic protease-activating factor 1 (Apaf-1), and procaspase-9, all of which interact in the cytosol. The downstream effector caspases, such as caspase-3, are activated after the apoptosome complex activates caspase-9. Active caspases execute the apoptotic process by cleaving various cellular substrates, leading to DNA fragmentation, cytoskeletal breakdown, and cell death (Tian et al. 2022; Vantaggiato et al. 2019).

9.11 Drp1 Activation Mitigates the Endogenous Antioxidant Defenses

Fragmented mitochondria, as a consequence of Drp1-induced fission, may exhibit impaired bioenergetics and altered ETC function. These dysfunctional mitochondria generate higher levels of ROS as byproducts of the ETC. ROS can react with GSH, a tripeptide composed of glutamate, cysteine, and glycine, through direct oxidation. ROS can directly transfer electrons to GSH, forming oxidized glutathione (GSSG) and other reactive intermediates (Zhou et al. 2021).

The reaction between ROS and GSH converts GSH to its oxidized form, GSSG. GSSG can be further reduced back to GSH through enzymatic reactions involving glutathione reductase and NADPH, ensuring a continuous supply of reduced GSH. However, under conditions of excessive ROS production or impaired antioxidant defense systems, the rate of GSSG formation may surpass the speed reductionist reduction rate, leading to a net depletion of cellular GSH levels. Protecting cells from oxidative assault alongside maintaining the cellular redox state in check are two of GSH's most essential functions. GSH acts as a potent reducing agent and participates in various antioxidant reactions, including detoxification. Depletion of GSH compromises these antioxidant functions, resulting in an impaired cellular antioxidant capacity. When GSH levels are low and antioxidant capacity is compromised, the body's ability to deal with ROS is compromised, leading to elevated oxidative stress. Damage to lipids, proteins, and DNA from increased ROS levels contributes to cellular malfunction, the development of neuronal death, and the progression of brain injury (Zitka et al. 2012; Presnell et al. 2013).

9.12 Conclusion

Understanding the pathogenic downstream cascade of various enzymes, such as NADPH oxidase, XO, COX, 5LOX, MMP9, and Drp1, in stroke pathogenesis is crucial for developing more targeted therapies. Neuronal injury and oxidative assault both get triggered by ROS generated through NADPH oxidase after a stroke. Stroke damage is mitigated when NADPH oxidase, or its components, is inhibited. This is because they prohibit ROS from being produced. Free radicals are produced by XO during purine metabolism. High levels of XO serve a role in the development of stroke pathology. Stroke damage and oxidative assault could be lessened by inhibiting XO. The neuronal injury and inflammation brought on by stroke are exacerbated by COX-mediated prostaglandin synthesis. Neuropathic pain and stroke damage can be mitigated with the help of selective COX inhibitors. Activated 5LOX generates leukotrienes, which contribute to inflammation. Stroke stimulates the expression of MMP9, a degrading enzyme of the ECM. Neuronal death, edema, and a breached BBB are all exacerbated by MMP9 activation. Stroke recovery may be enhanced by decreasing MMP9 activity. Due to its role in regulating mitochondrial fission, Drp1 dysregulation can lead to mitochondrial damage and ultimately neuronal death. Drp1 may mediate mitochondrial fission, which in turn increases mitochondrial fragmentation and energy failure after a stroke, may be by protecting mitochondrial function and reducing brain damage. Understanding the pathogenic downstream cascade of these enzymes is crucial to comprehending the harm caused by a stroke. Inhibiting these enzymes or the pathways they participate in can lessen the severity of strokes and increase outcomes for patients.

References

Berry CE, Hare JM (2004) Xanthine oxidoreductase and cardiovascular disease: molecular mechanisms and pathophysiological implications. J Physiol 555(3):589–606

Brilha S, Ong CW, Weksler B, Romero N, Couraud PO, Friedland JS (2017) Matrix metalloproteinase-9 activity and a downregulated Hedgehog pathway impair blood-brain barrier function in an in vitro model of CNS tuberculosis. Sci Rep 7(1):1–5

Candee R, Wilkenson R, Schreiber M, DeCenzo M (2023) The roles of neuroinflammation and glutamatergic excitotoxicity in treatment-resistant depression. JAAPA 36(4):12–17

Caron MM, Emans PJ, Sanen K, Surtel DA, Cremers A, Ophelders D, van Rhijn LW, Welting TJ (2016) The role of prostaglandins and COX-enzymes in chondrogenic differentiation of ATDC5 progenitor cells. PLoS One 11(4):e0153162

Chavda V, Chaurasia B, Garg K, Deora H, Umana GE, Palmisciano P, Scalia G, Lu B (2022) Molecular mechanisms of oxidative stress in stroke and cancer. Brain Disord 5:100029

Chemerinski E, Robinson RG (2000) The neuropsychiatry of stroke. Psychosomatics 41(1):5–14

Chung KF (1995) Leukotriene receptor antagonists and biosynthesis inhibitors: potential breakthrough in asthma therapy. Eur Respir J 8(7):1203–1213

Das SK, Banerjee TK, Biswas A, Roy T, Raut DK, Mukherjee CS, Chaudhuri A, Hazra A, Roy J (2007) A prospective community-based study of stroke in Kolkata, India. Stroke 38(3):906–910

Dong XX, Wang Y, Qin ZH (2009) Molecular mechanisms of excitotoxicity and their relevance to pathogenesis of neurodegenerative diseases. Acta Pharmacol Sin 30(4):379–387

Engblom D, Ek M, Saha S, Ericsson-Dahlstrand A, Jakobsson PJ, Blomqvist A (2002) Prostaglandins as inflammatory messengers across the blood-brain barrier. J Mol Med 80:5–15

Filip-Ciubotaru F, Manciuc C, Stoleru G, Liliana FO (2016) Nadph oxidase: structure and activation mechanisms (review). Note I. Med Surg J 120(1):29–33

Ghazanfari N, van Waarde A, Dierckx RA, Doorduin J, de Vries EF (2021) Is cyclooxygenase-1 involved in neuroinflammation? J Neurosci Res 99(11):2976–2998

Granger DN, Kvietys PR (2015) Reperfusion injury and reactive oxygen species: the evolution of a concept. Redox Biol 6:524–551

Guo C, Sun L, Chen X, Zhang D (2013) Oxidative stress, mitochondrial damage and neurodegenerative diseases. Neural Regen Res 8(21):2003

Hermann DM, Matter CM (2007) Tissue plasminogen activator–induced reperfusion injury after stroke revisited. Circulation 116(4):363–365

Hickey RW, Adelson PD, Johnnides MJ, Davis DS, Yu Z, Rose ME, Chang YF, Graham SH (2007) Cyclooxygenase-2 activity following traumatic brain injury in the developing rat. Pediatr Res 62(3):271–276

Higashi Y, Aratake T, Shimizu T, Shimizu S, Saito M (2021) Protective role of glutathione in the hippocampus after brain ischemia. Int J Mol Sci 22(15):7765

Hotter G, Closa D, Gelpi E, Prats N, Rosello-Catafau J (1995) Role of xanthine oxidase and eicosanoids in development of pancreatic ischemia-reperfusion injury. Inflammation 19:469–478

Kelly PJ, Murphy S, Coveney S, Purroy F, Lemmens R, Tsivgoulis G, Price C (2018) Anti-inflammatory approaches to ischaemic stroke prevention. J Neurol Neurosurg Psychiatry 89(2):211–218

Khandoga A, Kessler JS, Hanschen M, Khandoga AG, Burggraf D, Reichel C, Hamann GF, Enders G, Krombach F (2006) Matrix metalloproteinase-9 promotes neutrophil and T cell recruitment and migration in the postischemic liver. J Leukoc Biol 79(6):1295–1305

Kluger MA, Zahner G, Paust HJ, Schaper M, Magnus T, Panzer U, Stahl RA (2013) Leukocyte-derived MMP9 is crucial for the recruitment of proinflammatory macrophages in experimental glomerulonephritis. Kidney Int 83(5):865–877

Li C, Chen C, Qin H, Ao C, Chen J, Tan J, Zeng L (2022) The role of mitochondrial dynamin in stroke. Oxidative Med Cell Longev 2022:2504798

Loukogeorgakis SP, van den Berg MJ, Sofat R, Nitsch D, Charakida M, Haiyee BH, de Groot E, MacAllister RJ, Kuijpers TW, Deanfield JE (2010) Role of NADPH oxidase in endothelial ischemia/reperfusion injury in humans. Circulation 121(21):2310–2316

Malemud CJ (2017) Matrix metalloproteinases and synovial joint pathology. Prog Mol Biol Transl Sci 148:305–325

Mark LP, Prost RW, Ulmer JL, Smith MM, Daniels DL, Strottmann JM, Brown WD, Hacein-Bey L (2001) Pictorial review of glutamate excitotoxicity: fundamental concepts for neuroimaging. Am J Neuroradiol 22(10):1813–1824

Matsushima S, Tsutsui H, Sadoshima J (2014) Physiological and pathological functions of NADPH oxidases during myocardial ischemia–reperfusion. Trends Cardiovasc Med 24(5):202–205

Michaluk P, Kaczmarek L (2007) Matrix metalloproteinase-9 in glutamate-dependent adult brain function and dysfunction. Cell Death Differ 14(7):1255

Mohamed WS, Abd ElGawad EA, ElMotayam AS, Fathy SE (2019) Cardio embolic stroke and blood biomarkers: diagnosis and predictors of short-term outcome. Egypt J Neurol Psychiatry Neurosurg 55(1):1–9

Nuss P (2015) Anxiety disorders and GABA neurotransmission: a disturbance of modulation. Neuropsychiatr Dis Treat 11:165–175

Parascandolo A, Laukkanen MO (2019) Carcinogenesis and reactive oxygen species signaling: interaction of the NADPH oxidase NOX1–5 and superoxide dismutase 1–3 signal transduction pathways. Antioxid Redox Signal 30(3):443–486

Pegadraju H, Thomas JA, Kumar R (2022) Mechanistic and therapeutic role of Drp1 in the pathogenesis of stroke. Gene 855:147130

Popp A, Urbach A, Witte OW, Frahm C (2009) Adult and embryonic GAD transcripts are spatiotemporally regulated during postnatal development in the rat brain. PLoS One 4(2):e4371

Presnell E, Bhatti G, Numan L, Lerche M, Alkhateeb S, Ghalib M, Shammaa M, Kavdia M (2013) Computational insights into the role of glutathione in oxidative stress. Curr Neurovasc Res 10(2):185–194

Radi R (2018) Oxygen radicals, nitric oxide, and peroxynitrite: redox pathways in molecular medicine. Proc Natl Acad Sci 115(23):5839–5848

Rådmark O, Werz O, Steinhilber D, Samuelsson B (2015) 5-Lipoxygenase, a key enzyme for leukotriene biosynthesis in health and disease. Biochim Biophys Acta 1851(4):331–339

Ramos-DeSimone N, Hahn-Dantona E, Sipley J, Nagase H, French DL, Quigley JP (1999) Activation of matrix metalloproteinase-9 (MMP-9) via a converging plasmin/stromelysin-1 cascade enhances tumor cell invasion. J Biol Chem 274(19):13066–13076

Shirley R, Ord EN, Work LM (2014) Oxidative stress and the use of antioxidants in stroke. Antioxidants 3(3):472–501

Suarez I, Bodega G, Fernandez B (2002) Glutamine synthetase in brain: effect of ammonia. Neurochem Int 41(2-3):123–142

Szklarczyk A, Lapinska J, Rylski M, McKay RD, Kaczmarek L (2002) Matrix metalloproteinase-9 undergoes expression and activation during dendritic remodeling in adult hippocampus. J Neurosci 22(3):920–930

Takeda N, O'Dea EL, Doedens A, Kim JW, Weidemann A, Stockmann C, Asagiri M, Simon MC, Hoffmann A, Johnson RS (2010) Differential activation and antagonistic function of HIF-α isoforms in macrophages are essential for NO homeostasis. Genes Dev 24(5):491–501

Tavazzani E, Tritto S, Spaiardi P, Botta L, Manca M, Prigioni I, Masetto S, Russo G (2014) Glutamic acid decarboxylase 67 expression by a distinct population of mouse vestibular supporting cells. Front Cell Neurosci 8:428

Tian H, Chen X, Liao J, Yang T, Cheng S, Mei Z, Ge J (2022) Mitochondrial quality control in stroke: from the mechanisms to therapeutic potentials. J Cell Mol Med 26(4):1000–1012

Tilleux S, Hermans E (2007) Neuroinflammation and regulation of glial glutamate uptake in neurological disorders. J Neurosci Res 85(10):2059–2070
Turner RJ, Sharp FR (2016) Implications of MMP9 for blood brain barrier disruption and hemorrhagic transformation following ischemic stroke. Front Cell Neurosci 10:56
Vantaggiato C, Castelli M, Giovarelli M, Orso G, Bassi MT, Clementi E, De Palma C (2019) The fine tuning of Drp1-dependent mitochondrial remodeling and autophagy controls neuronal differentiation. Front Cell Neurosci 13:120
Vongsfak J, Pratchayasakul W, Apaijai N, Vaniyapong T, Chattipakorn N, Chattipakorn SC (2021) The alterations in mitochondrial dynamics following cerebral ischemia/reperfusion injury. Antioxidants 10(9):1384
Wang ML, Huang XJ, Fang SH, Yuan YM, Zhang WP, Lu YB, Ding Q, Wei EQ (2006) Leukotriene D4 induces brain edema and enhances CysLT2 receptor-mediated aquaporin 4 expression. Biochem Biophys Res Commun 350(2):399–404
Wang Y, Yang Y, Zhang S, Li C, Zhang L (2020) Modulation of neuroinflammation by cysteinyl leukotriene 1 and 2 receptors: implications for cerebral ischemia and neurodegenerative diseases. Neurobiol Aging 87:10
Zeng MY, Miralda I, Armstrong CL, Uriarte SM, Bagaitkar J (2019) The roles of NADPH oxidase in modulating neutrophil effector responses. Mol Oral Microbiol 34(2):27–38
Zhou X, Chen H, Wang L, Lenahan C, Lian L, Ou Y, He Y (2021) Mitochondrial dynamics: a potential therapeutic target for ischemic stroke. Front Aging Neurosci 13:721428
Zitka O, Skalickova S, Gumulec J, Masarik M, Adam V, Hubalek J, Trnkova L, Kruseova J, Eckschlager T, Kizek R (2012) Redox status expressed as GSH: GSSG ratio as a marker for oxidative stress in paediatric tumour patients. Oncol Lett 4(6):1247–1253

Future Directions and Challenges in Enzyme Research for Neurodegenerative Diseases

10

Karishma Manohar Rathi, Priyanka Vilas Pingat, and Sayed Mohammed Firdous

Abstract

Neurodegenerative diseases (NDs) pose a significant health challenge characterized by progressive neuronal loss and cognitive decline. Despite extensive research, effective therapeutic interventions remain elusive, necessitating the exploration of innovative approaches. Enzymes, with their specificity and catalytic prowess, emerge as promising candidates for ND treatment. This chapter reviews the current landscape of enzyme research in NDs, emphasizing key areas like protein aggregation modulation, neuroinflammation control, and metabolic pathway repair. The discussion also delves into challenges hindering clinical translation, including the intricate nature of NDs, the absence of reliable biomarkers, protein misfolding complexities, and the need for multitarget therapeutic strategies. Moreover, challenges such as the development of early diagnostic tools and overcoming blood-brain barrier limitations are explored. Future research directions are proposed, encompassing strategies like targeting protein aggregation, modulating protein-folding pathways, computational modeling, enzyme design, delivery and targeting strategies, protein and gene delivery systems, nanozymes, molecular imaging, stem cell therapy, and novel technologies in enzyme research. Addressing these challenges and leveraging the unique potential of enzymes may usher in a promising era for ND treatment.

K. M. Rathi · P. V. Pingat
Department of Pharmacy Practice, Dr. D.Y. Patil Institute of Pharmaceutical Sciences and Research, Pimpri-Chinchwad, India

S. M. Firdous (✉)
Department of Pharmacology, Calcutta Institute of Pharmaceutical Technology & AHS, Howrah, West Bengal, India

B. Mathew, D. G. T. Parambi (eds.), *Enzymes in Neurodegenerative Disorders*,
https://doi.org/10.1007/978-981-97-6822-6_10

Keywords

Neurodegenerative diseases · Enzymes · Protein aggregation · Neuroinflammation · Metabolic dysfunction · Therapeutic challenges · Future directions

10.1 Introduction

Neurodegenerative diseases (NDs) such as Alzheimer's disease (AD), Parkinson's disease (PD), and amyotrophic lateral sclerosis (ALS) are marked by a gradual decline in neurons, resulting in cognitive deterioration, impaired motor function, and eventual mortality. Despite significant research efforts, current therapeutic options primarily manage symptoms rather than addressing the underlying disease pathology. This necessitates exploring novel therapeutic strategies, and enzymes, with their remarkable catalytic power and substrate specificity, emerge as promising candidates (Bar-Or et al. 2011).

10.1.1 Current Applications of Enzymes in ND Research

10.1.1.1 Protein Aggregation Modulation

A characteristic feature observed in numerous NDs is the anomalous buildup of misfolded protein aggregates, exemplified by amyloid-β (Aβ) in AD and α-synuclein in PD. Enzymes can be targeted toward these aggregates for degradation or disaggregation. For example, lysosomal enzymes like glucocerebrosidase have shown promise in Gaucher's disease, while engineered proteases like neprilysin are being explored for Aβ clearance in AD (Soto and Pritzkow 2018).

10.1.1.2 Neuroinflammation Control

The advancement of ND s relies significantly on neuroinflammation. Investigations into enzymes such as papain and bromelain reveal their anti-inflammatory properties, suggesting potential applications in alleviating neuroinflammation in Parkinson's and ADs (Zhang et al. 2023).

10.1.1.3 Metabolic Pathway Repair

Dysfunctional metabolic pathways contribute to ND pathogenesis. Enzymes can correct these imbalances by replenishing deficient metabolites or breaking down toxic ones. For example, the enzyme glutamate decarboxylase is a prospective ALS treatment, aiming to restore glutamate homeostasis (Moos et al. 2021).

10.2 Challenges in Enzyme Research for ND Diseases

10.2.1 Complexity of NDs

The complexity of NDs poses a significant challenge in formulating universal therapeutic strategies. These diseases can be classified by primary clinical features (e.g., PD, neuronal disease, and dementia), anatomical distribution (such as frontal temporal degeneration, extrapyramidal disorders, or spinocerebellar degeneration), or primary molecular abnormalities. Amyloidoses, Tauopathies, β-synucleinopathies, and TDP-43 proteinopathies are among the most prevalent NDs. Although specific protein accumulation and anatomical vulnerability generally define these diseases, they share fundamental processes associated with neurodegenerative dysfunctions and deaths, including proteotoxic stress and its effects on ubiquitin–proteasome and autophagosome/lysosome systems, oxidative stress, programmed cell death, and neuroinflammation. The different onset stages, progression rates, and overlaps in pathologies contribute to the complexity of these diseases (Dugger and Dickson 2017; Noble and Burns 2010).

10.2.2 Lack of Reliable Biomarkers

A crucial aspect of the mechanisms underlying NDs often involves the excessive production of precursor proteins linked to the creation of abnormal proteins characterized by a structure that lacks stability. These modified proteins tend to assemble and amass both inside and outside cells, forming protein aggregates potentially connected to the decline of neurons in the affected brain area. These clusters are identified as biomarkers (Basha et al. 2021). Biomarkers can be classified according to their intended purpose. The first group includes diagnostic biomarkers, including preclinical markers for early detection of neurodegenerative processes, prodromic markers for symptoms without clear disease classification, and clinical markers for differential diagnosis. The second group consists of prognostic biomarkers, estimates of disease risk, and predicts progression rates. The third group includes biomarkers that identify phases or phases of disease. Finally, biomarkers for therapeutic purposes are employed to observe the reaction to the intervention as shown in Fig. 10.1 (Cova and Priori 2018).

The identification of biomarkers is currently an important subject in the field of NDs. By definition, biological indicators are objectively measurable and evaluated indicators that describe the physiological processes, pathological processes, and pharmaceutical reactions to therapeutic interventions. The main purpose of biomarkers is to improve clinical diagnosis to improve differential diagnosis accuracy. Biomarkers are intended to help estimate stages of disease and progression and reflect treatment response rates. The sensitivity and specificity of biomarkers are important differences. Sensitivity reflects the success rate of detecting the presence of the condition observed in the subject. The specificity reflects the ability to select the cases of the study that do not occur exactly (Koníčková et al. 2022; Hansson 2021).

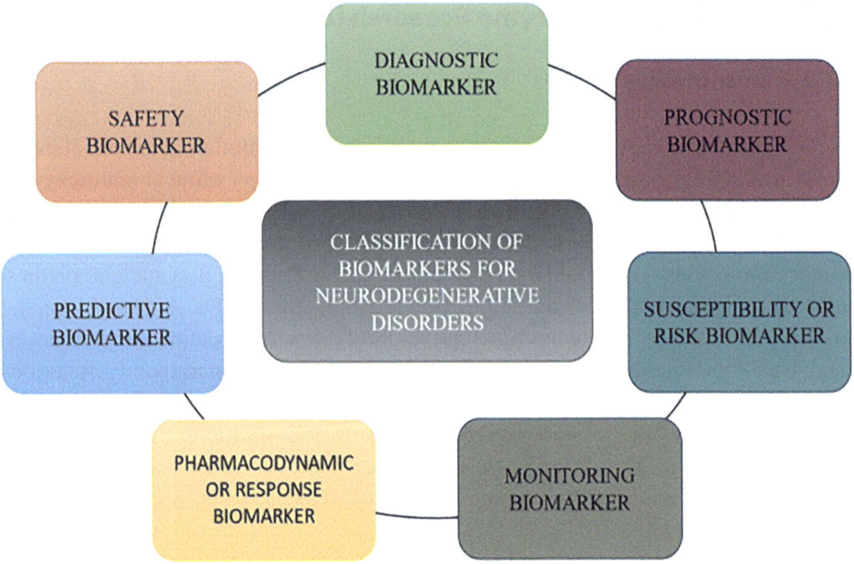

Fig. 10.1 Classification of biomarkers for neurodegenerative diseases

The lack of reliable and specific biomarkers for the diagnosis of neurodegenerative conditions poses a significant hurdle in developing effective treatments and monitoring disease progression (Swift et al. 2021). NDs exhibit a complex interplay of various molecular signaling pathways, which can be examined through bodily fluids, particularly cerebrospinal fluid (CSF) and blood (Barkovits et al. 2020). CSF, being a valuable fluid, mirrors the biochemical activities in brain tissue, rendering it a suitable matrix for analyzing the central nervous system (CNS). Although this allows for early diagnosis of preclinical diseases, less invasive plasma or serum plasma is preferred for monitoring NDs (Swift et al. 2021; Gaetani et al. 2020).

Communication between the blood and the brain involves complex processes mediated by the blood-brain barrier (BBB), lymphatic vessels, and the glymphatic system. The analysis of blood for CNS biomarkers encounters challenges:

1. CNS biomarkers may undergo dilution in the blood due to low concentrations in the CSF.
2. Lack of CNS specificity in biomarkers expressed in peripheral tissues may overshadow CNS-related information.
3. Proteins like albumins, globulins, transferrin, etc., in the blood pose analytical challenges due to potential interference.
4. Heterophilic antibodies present in the bloodstream may disrupt immune assays.
5. Proteolytic degradation in the plasma complicates accurate measurements of observed analytes.
6. Pre-analytical changes underscore the need to consider factors impacting analytical result reliability (Zetterberg and Burnham 2019).

Currently, the detection of ND heavily relies on brain imaging methods focusing on functionality or intrusive measures like lumbar puncture, which are costly and not universally available (Noble and Burns 2010). It is essential to identify early signs of NDs and consider the key role of early intervention in the effectiveness of therapy to address the disease (Luebke et al. 2023). However, the intricate nature of NDs, involving multiple proteins, often accompanied by neuroinflammation and exhibiting diverse ages of onset and progression rates, poses challenges in identifying specific biomarkers (MDPI 2024). Additionally, the overlap in pathological features among different diseases and the concurrent occurrence of two or more NDs further complicates the quest for dependable biomarkers (Noble and Burns 2010).

The utilization of proteomic analysis is on the rise as an impartial approach to identifying disease markers. When combined with machine learning algorithms, this screening method can reveal new biomarkers for NDs. Nonetheless, the translation of these discoveries into clinical practice necessitates additional validation and standardization efforts (He et al. 2018).

10.2.3 Understanding Misfolded Proteins

Understanding misfolded proteins poses a considerable challenge in the realm of NDs. Protein misfolding can give rise to the buildup of toxic aggregates, a common feature in multiple ND s such as AD, PD, Huntington's, and ALS. Genetic mutations or environmental influences have the potential to instigate protein misfolding, disturbing regular cellular processes, and, ultimately, culminating in neuronal impairment and cell demise (Reynaud 2010; Sweeney et al. 2017).

The treatment of misfolding of proteins poses unique challenges in the field of drug discovery and development. These challenges are due to factors such as (1) the dynamic nature of the proteins involved, (2) uncertainty about the specific forms of the disease protein (monomer, oligomer, or insoluble aggregate) that primarily contribute to cell toxicity, (3) limited understanding of the components of the cell proteostatic machinery that interact with these disease proteins, and (4) the lack of established clinically tested biomarkers (Sweeney et al. 2017).

Cells are frequently confronted with an ongoing influx of defective proteins due to biogenesis errors, disease-related mutations, and physiological stress factors (see Table 10.1 for a compilation of misfolded proteins linked to NDs) (Sweeney et al. 2017).

Exploring protein interaction networks in NDs has yielded insights into the mechanisms governing protein misfolding and aggregation. Mutations in genes encoding proteins prone to aggregation are prevalent in rare familial NDs (Bertram and Tanzi 2005). While the precise disease mechanisms remain incompletely understood, research spanning the past two decades has unveiled a substantial protein-protein interaction network that regulates the aggregation of key proteins associated with neurodegeneration. Proteomic investigations have uncovered an extensive network of protein-protein interactions pivotal to the pathogenesis of NDs. These

Table 10.1 Misfolded proteins associated with ND

Proteinopathy	Aggregating protein(s)
Parkinson's disorder	α-synuclein
Alzheimer's disorder	Amyloid beta (Aβ) peptide; Tau
Huntington's disorder	Huntingtin with tandem glutamine repeats
Multiple tauopathies	Tau protein (microtubule-associated)
Amyotrophic lateral sclerosis	Superoxide dismutase 1
Familial amyloidotic polyneuropathy	Transthyretin (mutant forms)
Spongiform encephalopathies	Prion proteins

interactions involve crucial proteins prone to aggregation, including amyloid beta, tau, alpha-synuclein, and TDP-43, among others (Calabrese et al. 2022).

Understanding the interplay between diverse proteins and pathways in NDs is essential for devising targeted therapeutic interventions capable of modulating the activity of misfolded proteins and impeding the formation of toxic aggregates. Ongoing research on therapeutic options focuses on various stages in the generation and aggregation of misfolded proteins, encompassing synthesis, chaperone-assisted folding and transport, and clearance (Kory et al. 2021).

10.2.4 Multitarget Therapeutic Approaches

Effective treatment for NDs, known for their multifactorial nature involving diverse proteins and pathways, necessitates multitargeted therapeutic approaches. The identification and targeting of specific events crucial for rescuing afflicted neurons are vital components in the development of such treatments (Noble and Burns 2010; Dokholyan et al. 2022; NanoTemper Technologies 2020).

A significant challenge in enzyme research for NDs is the absence of disease-modifying therapies. Presently, therapies for most neurodegenerative conditions are symptomatic, with limited availability of disease-modifying strategies (Noble and Burns 2010). Recognizing the complex nature of ND, the creation of therapeutic approaches targeting multiple aspects becomes essential. These approaches can concurrently address multiple pathways and proteins integral to the disease process (Dokholyan et al. 2022; NanoTemper Technologies 2020).

The complex nature of NDs encompasses diverse biological processes, such as neuroinflammation, oxidative stress, and mitochondrial dysfunctions. All of these factors collectively contribute to the development and pathogenesis of these conditions, as illustrated in Fig. 10.2 (Bolognesi et al. 2009; Bhat et al. 2015). Given this complexity, targeting a single pathway or protein may prove insufficient, emphasizing the need for interventions capable of simultaneously modulating multiple targets. The challenge in drug discovery and development lies in acquiring a comprehensive understanding of the molecular mechanisms that underlie diseases and identifying specific targets for rescuing afflicted neurons (Ibrahim and Gabr

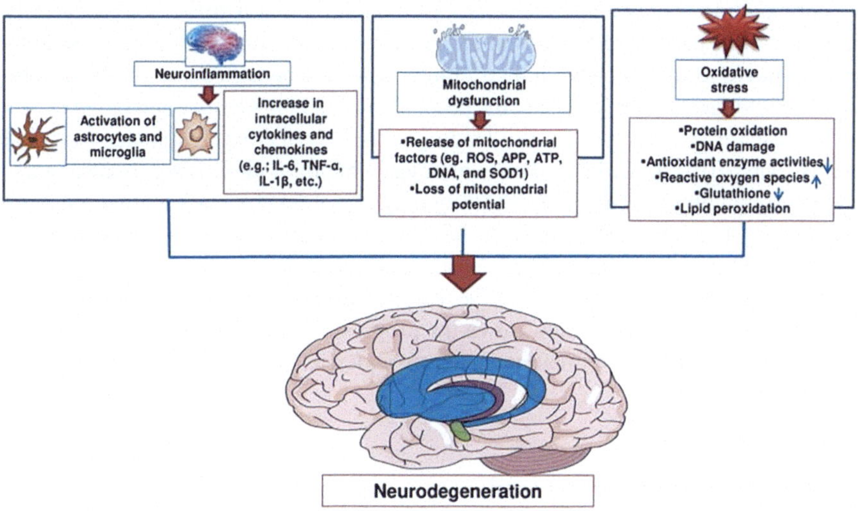

Fig. 10.2 Various biological processes, such as oxidative stress, neuroinflammation, and mitochondrial dysfunctions, contribute to the advancement and pathogenesis of neurodegenerative disease

2019). Furthermore, translating this knowledge into the creation of effective multi-target therapies remains a significant challenge in the neurodegeneration field (Dokholyan et al. 2022).

Several studies have focused on developing multitargeted therapeutic strategies for neurodegenerative conditions that encompass AD and PDs. A 2020 study published in NCBI proposed a multitarget strategy for treating AD, targeting multiple pathways involved in the disease process (Maramai et al. 2020). Another study from 2013, published in ScienceDirect, explored the application of multitarget-directed ligands for addressing NDs (Bolognesi et al. 2009). Additionally, a recent editorial from 2023, published in Frontiers in Neuroscience, examined the hurdles and possibilities associated with medical trials featuring GLP-1 receptor agonists in the context of ND (Tikhonova et al. 2023).

10.2.5 Developing Early Diagnostic Tools

The creation of diagnostic tools for the early detection of NDs is a major challenge in enzyme research. These diseases are multifactorial, involving multiple proteins, often accompanied by neuroinflammation, and differing in age of onset and rate of progression, making it difficult to identify specific and reliable biomarkers. The lack of reliable and specific biomarkers for rapid diagnosis of NDs hinders the advancement of successful treatments and monitoring of disease progression (Doroszkiewicz et al. 2022).

For example, AD is a neurodegenerative condition that is difficult to accurately diagnose in its early stages. Current diagnostic methods rely on clinical symptoms and cognitive tests, which are not specific or sensitive enough to detect the disease in its early stages. Identifying specific biomarkers can help identify and target events in the disease process, potentially leading to more effective treatments (Agrawal and Biswas 2015). In tackling these obstacles, a blend of identifying disease markers, neuroimaging, and diagnostic testing may reveal ND biomarkers that are more sensitive and specific. This, in turn, has the potential to lay the groundwork for personalized treatment approaches (Noble and Burns 2010).

The development of reliable and specific biomarkers is essential to fast diagnosis and disease surveillance to slow down the progress of neurodegenerative processes and to provide effective treatments (Doroszkiewicz et al. 2022; Agrawal and Biswas 2015).

10.2.6 Limitations of the BBB

The BBB acts as an innate protective barrier, protecting the CNS. It is effective against toxins and pathogens circulating in the bloodstream (Wu et al. 2023). Comprising microvascular endothelial cells along brain capillaries, the BBB establishes a metabolic barrier through the expression of various enzymes that alter both internal and external molecules (Zlokovic 2008). While this barrier plays a crucial role in CNS protection, it presents difficulties in delivering therapeutics to the brain and hinders the progress of effective treatments for NDs.

A major limitation arises from the role of the BBB in interfering with the transport of therapeutic agents to the brain, thereby impeding progress in enzyme research for NDs (Ebrahimi et al. 2020). Moreover, under normal physiological conditions, the BBB is not a static entity and undergoes alterations in NDs. External factors such as pro-inflammatory cytokines, nitric oxide, and neurosteroids can impair BBB function (Carvey et al. 2009).

To address these challenges, scientists have considered various strategies to overcome the limitations of the BBB. These methods vary from interfering with tight junctions to enhancing transcytosis through peptide-based delivery systems, local delivery approaches, nanoparticle delivery, and cell-based delivery. Despite these efforts, additional research and development are essential for enhancing the effectiveness of these strategies in treating NDs (Ebrahimi et al. 2020; Kadry et al. 2020).

10.3 Future Prospectives in Enzyme Research for ND Diseases: A Glimpse Into Hope

Enzymes, nature's exquisite protein catalysts, offer immense potential in tackling the proteinopathy hallmark of NDs. Their remarkable specificity and efficiency in accelerating biochemical reactions make them ideal candidates for therapeutic intervention. Here, we delve into the exciting future directions of enzyme research for NDs, exploring novel strategies and promising avenues for development.

10.3.1 Targeting Protein Aggregation

10.3.1.1 Proteases and Peptidases for Degrading Misfolded Proteins

Proteases and peptidases, enzymes with the ability to break down proteins into smaller peptides or amino acids, play a crucial role in degrading misfolded proteins. Misfolded proteins, characterized by incorrect folding, can aggregate and form toxic clumps implicated in the development of NDs such as AD, PD, and ALS.

A specific category of protease, papain-based enzymes, demonstrates efficacy in breaking down alpha-synuclein aggregates. Alpha-synuclein is a protein commonly found in Lewy's body and is a characteristic of PD. Studies indicate that papain-based enzymes can effectively degrade alpha-synuclein aggregates in both animal and cellular models of PD (Gandhi et al. 2019; Chien Lab 2024).

10.3.1.2 Chaperones Facilitating Protein Folding

Chaperones are proteins that help other proteins fold correctly and prevent them from misfolding and play a pivotal role in cellular function. Heat shock proteins, a subtype of chaperones, are particularly significant for cellular stress protection. Research has shown that heat shock protein helps prevent the accumulation of misfolded proteins in cells and animal models of ND (Ross and Poirier 2021; Waheed et al. 2022) as shown in (Fig. 10.3).

An encouraging path in ND treatment includes employing proteases and peptidases to degrade misfolded proteins, complemented by chaperones facilitating correct protein folding. This innovative approach holds the potential to prevent the formation of toxic protein aggregates and enhance neuronal function (Cuny 2012).

10.3.2 Modulating Protein Folding Pathways: Isomerases and Foldases as ND Treatment

The future landscape of enzyme research for NDs, particularly concerning the modulation of protein folding pathways, holds considerable promise. Isomerases and foldases emerge as noteworthy contenders, demonstrating potential in catalyzing the correct folding of proteins implicated in diseases like α-synuclein in PD and tau in AD. This catalytic action aims to prevent protein aggregation and ensuing neurotoxicity (Cuny 2012; Cortez and Sim 2014).

In parallel, exploring the unfolded protein response (UPR) and the therapeutic potential of curcumin to reduce protein misfolding and aggregation represents a promising avenue for future research (Mukherjee et al. 2021).

Moreover, strides in understanding protein interaction networks within NDs provide valuable insights into the mechanisms governing protein misfolding and aggregation. These insights present new prospects for developing targeted therapeutic

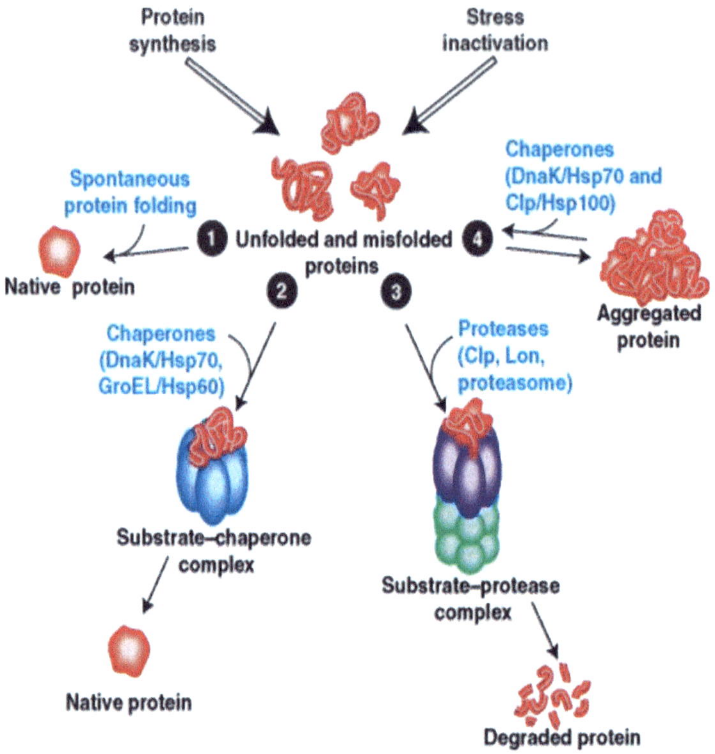

Fig. 10.3 Chaperone-assisted protein folding

interventions (Calabrese et al. 2022). The investigation into chemical chaperones, osmolytes, and UPR, coupled with the advancement of engineered enzymes, marks an exciting direction for the field. This multifaceted approach holds significant potential to profoundly influence the management and treatment of NDs (Sweeney et al. 2017).

10.3.3 Computational Modelling and Enzyme Design

Recent strides in the realm of ND research underscore the paramount importance of computational modeling and in silico methodologies, particularly in the domain of enzyme design. In 2021, Morales-Navarro et al. conducted a study that emphasized the integration of theoretical and experimental methodologies in the design of drugs, with a particular focus on addressing NDs. This comprehensive approach demonstrated the potential efficacy of computational modeling in discerning novel therapeutic strategies (Morales-Navarro et al. 2019).

In addition, the widespread use of genome-scale metabolic models has proven instrumental in uncovering metabolic signatures within NDs, offering invaluable insights into disease pathophysiology and revealing potential targets for therapeutic intervention (Baloni et al. 2021).

Anticipating the future, the incorporation of in silico methodologies, including virtual screening of enzyme libraries and targeted modifications, holds tremendous promise for crafting enzymes that are highly specific and efficient in treating NDs. Through the utilization of computational modeling and simulations, researchers aim to deepen their comprehension of disease-specific metabolic dysregulations and pinpoint crucial biological processes underpinning NDs. This acquired knowledge serves as a foundation for the design of novel enzymes endowed with enhanced therapeutic potential.

Comparing recent advancements with future prospects, the trajectory of in silico approaches in enzyme research for NDs is marked by an escalating emphasis on precision and specificity. Leveraging computational tools, researchers are diligently working toward developing enzymes tailored to address the distinctive metabolic dysregulations associated with diverse NDs. This nuanced strategy can bring about a revolution in the field, opening avenues for the creation of highly targeted and efficient enzyme-based therapies for these incapacitating conditions (Cuny 2012).

10.3.4 Enzyme Delivery and Targeting Strategies

10.3.4.1 Delivery Systems for Proteins and Genes in the Management of NDs

Exploring protein and gene delivery systems to address diverse diseases is a captivating yet intricate and resource-demanding area of research. In recent times, there has been a growing fascination with the utilization of genes and recombinant proteins like growth factors and monoclonal antibodies, owing to their pharmacological potential in treating genetic and metabolic conditions (Kaiafas et al. 2020; Miliotou et al. 2021; Manoutcharian et al. 2017).

Despite substantial progress in this field, there remains a need for further investigation into early diagnosis and successful treatment. Traditional medications for NDs have historically focused on symptom relief rather than targeting the root cause. Treatment options focused on genes and proteins are mainly directed toward the CNS, the eye, or the brain. However, their effectiveness is compromised by difficulties in crossing important physiological barriers such as the BBB, blood-cerebrospinal fluid barrier (BCSFB), and blood-retinal barrier (BRB) (Chen et al. 2020; Dunbar et al. 2018; Siafaka et al. 2021).

The current demand underscores the necessity to explore novel drug agents, emphasizing biological molecules with reduced toxicity and heightened safety profiles for NDs. Additionally, identifying innovative delivery routes, such as intranasal administration or subpial injections, becomes crucial to improve the penetration of the BBB or BCSFB (Wu et al. 2021; Miyanohara et al. 2016).

Traditional routes of administration are often unsuitable for biomolecules due to their limited half-life and susceptibility to degradation by gastric juices, resulting in reduced therapeutic efficacy (Zhao et al. 2012). Intravenous administration poses a challenge in crossing the BBB, but intramuscular or subcutaneous administration offers an alternative for efficiently traversing physiological barriers and gaining clinical benefits (Bittner et al. 2018). Consequently, present and future therapeutic approaches for NDs should integrate disease management with successful barrier penetration to mitigate potential side effects.

Although current clinical trial data supporting the application of gene and protein delivery in the treatment of NDs are insufficient, the trajectory of medicine appears to favor such molecules. This inclination arises from the inadequacies of conventional therapies, which often yield limited results. The primary challenges associated with these molecules include their half-life, potential degradation, and the risk of adverse effects (Siafaka et al. 2022).

10.3.4.2 Nanozymes: Next-Generation Artificial Enzymes

Nanozymes, possessing enzyme-like characteristics, represent a cutting-edge approach designed to address the constraints inherent in natural enzymes and conventional artificial enzyme counterparts. Despite considerable advancements, there exists ample opportunity for furthering the research on nanozymes, with key areas of focus outlined (Wu et al. 2019):

- Rational design of nanozymes: It is important to combine experimental and computational studies to advance our understanding of the catalytic mechanisms of nanozymes. Integrating concepts such as single-atom nanozymes (active sites) and single-nanozyme catalysis enable the creation of nanozymes with tailored characteristics for optimal performance.
- Sabatier principle and explanation of nanozyme activity: When following the Sabatier principle, identification of optimal catalytic interactions is of paramount importance. From an electronic point of view, activity descriptors pave the way to establishing general design rules for high-performance nanozymes (Medford et al. 2015).
- Mimicking of protein scaffolds and the native microenvironment: Given the important role that protein scaffolds and the enzyme's native microenvironment play in selectivity and efficiency, future nanozyme research will explore new strategies to address these challenges (Parton et al. 1994).
- Expanding the range of nanozymes and beyond: To cover all essential enzymatic reactions, nanozymes can be developed by analogy with the active sites of natural enzymes and incorporating them into metal–organic frameworks or other nanomaterials. Future studies should investigate the generation of organelles and cell mimics (Wang et al. 2016).
- Specific nanozymes and nanozymes that mimic multiple enzymes: Addressing the challenge of low substrate specificity is crucial. Introducing recognition mechanisms, such as the creation of protein-like (or aptamer-like) binding pockets or molecular imprinting of selective substrate pockets, can lead to nanozymes demonstrating enhanced specificity (Hu et al. 2018).
- Multifunctional nanozymes: Utilizing the physicochemical attributes of nanomaterials creates possibilities for crafting multifunctional nanozymes. These

nanozymes may possess magnetic, optical, and thermal features for precise sensing, sustainable chemistry, and combination therapy (Wu et al. 2018).

- Biological effects of nanozymes: Systematically evaluating the benefits and risks of nanozymes, including cellular fate, clinical toxicity, pharmacokinetics, immunogenicity, and long-term environmental effects, is essential for translational applications.
- Standardization of nanozyme research: Establishing standards for quantitatively determining catalytic activities is crucial. The catalytic activities of nanomaterials are significantly influenced by their distinct characteristic, the specific surface area.
- Killer Applications: While Nanozymes hold transformative potential across diverse domains, securing enduring support necessitates their identification and application in distinctive niches.

In the realm of NDs, nanozymes show potential for addressing conditions like AD and PD through their ability to target enzymes involved in amyloid degradation and other disease-related enzymatic processes (Turner 2004). Furthermore, nanozymes can contribute to disease imaging and detection, offering a cost-effective and operationally simple strategy for live cell and organelle imaging (Wang et al. 2020).

10.3.5 Future Directions in Molecular Imaging of NDs

Significant advances in molecular imaging techniques have revolutionized the comprehension and diagnosis of NDs. The conventional approach, defining and diagnosing based on clinical phenotypes, has not yielded breakthroughs in curative or disease-modifying treatments. Notably, AD introduced a new diagnostic concept centered around biomarker testing for amyloid, tau, and neurodegeneration, culminating in the approval of the β-amyloid-reducing drug aducanumab by the Food and Drug Administration (Jack et al. 2018; Garibotto et al. 2021).

A pivotal shift in defining NDs from a syndromal construct to a biological construct is anticipated. This transformation necessitates biomarker-based diagnoses, with positron emission tomography (PET) imaging potentially playing a decisive role in this paradigm shift (as shown in Fig. 10.4) (Weber et al. 2020; Brettschneider et al. 2015).

Magnetic resonance imaging (MRI) and PET stand out as the most extensively employed neuroimaging methods in the context of NDs. Their pivotal functions encompass the identification, monitoring, diagnosis, and assessment of treatment for neurological diseases. In particular, PET enables the imaging of protein aggregates like β-amyloid, α-synuclein, tau, and TDP-43, thus facilitating early and accurate diagnosis of NDs. This capability has spurred the development of novel tracers for protein aggregation pathology imaging, significantly advancing the early and accurate diagnosis of these disorders (Risacher and Saykin 2021; Palaninathan et al. 2015).

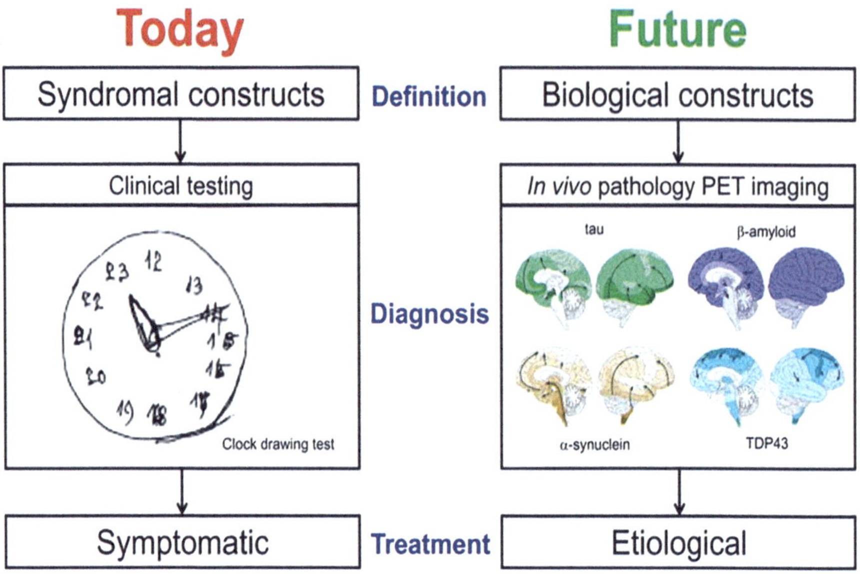

Fig. 10.4 Presenting the anticipated paradigm shift in the definition, diagnosis, and treatment of neurodegenerative diseases, highlighting the role of positron emission tomography imaging

Beyond protein aggregation pathology imaging, molecular imaging techniques are evolving to encompass neurotransmitter systems beyond the currently explored dopaminergic, cholinergic, and serotonergic systems. This expansion includes imaging systems like the glutamate and opioid systems, offering a more comprehensive understanding of NDs. Technological advancements, such as dedicated brain PET, hybrid PET/MRI, and whole-body PET scanners, alongside refined image acquisition and processing protocols, will further augment the capabilities of molecular imaging. These advancements hold the potential for wider and more effective applications in clinical settings, as well as opening up new avenues for research applications (Risacher and Saykin 2021; Barthel et al. 2022).

The future of molecular imaging in NDs holds immense promise. Ongoing development of novel tracers for imaging protein aggregation pathology, progress in imaging neurotransmitter systems, and enhancements in imaging technology will fortify early and accurate diagnoses. These developments will not just enhance our comprehension of the fundamental pathophysiology of these disorders but also lay the groundwork for more efficient treatments and creative research applications (Barthel et al. 2022).

10.3.6 Stem Cell Therapy for NDs

Stem cell therapy exhibits the potential to address NDs, particularly when faced with the CNS's constrained self-healing ability. While certain brain regions house populations of native neural stem cells, the introduction of induced pluripotent stem cells (iPSCs) marks a breakthrough in this domain. The revelation that fibroblasts can undergo reprogramming to pluripotency and subsequently differentiate into distinct neuronal subtypes holds substantial implications for the field.

The recent emergence of iPSCs has opened the opportunity to study patient-specific disease mechanisms and progression by obtaining fibroblasts from skin biopsies of patients with NDs. This approach has the potential to pave the way for individually tailored therapeutic interventions. While technical challenges persist in this research domain, the manipulation of iPSCs emerges as a potent and innovative strategy for addressing NDs (Noble and Burns 2010).

10.3.7 Novel Technologies in Enzyme Research

Technological advancements, encompassing robotic microscopy, machine-learning approaches, CRISPR screening, and high-throughput screening methods have played pivotal roles in managing the substantial data generated by screens in ND research. Despite the capacity to amass extensive datasets, the translation of this wealth of information into effective diagnostic and therapeutic solutions for NDs remains elusive.

The application of these technologies has enabled the capture of vast amounts of data, but a comprehensive understanding of the intricate molecular mechanisms underlying these pathways is still lacking. Further studies are imperative to unravel the complexities of these molecular pathways and bridge the gap between data accumulation and tangible diagnostic and therapeutic breakthroughs (Kumar et al. 2022; Xu et al. 2024).

10.4 Conclusion

In conclusion, exploring enzymes as therapeutic entities for NDs represents a cutting-edge avenue with profound potential. The multifaceted approach, encompassing protein aggregation modulation, neuroinflammation control, and metabolic pathway repair, showcases the versatility of enzymatic interventions. Despite the promise, the translation of enzyme-based therapies faces obstacles rooted in the complexity of NDs, the absence of definitive biomarkers, and challenges associated with protein misfolding and BBB penetration. Overcoming these hurdles necessitates concerted efforts in developing multitargeted therapeutic strategies, early diagnostic tools, and innovative delivery systems. The envisioned future directions in research, spanning protein folding pathways, computational modeling, nanozymes, molecular imaging, and stem cell therapy, provide a roadmap for advancing ND treatment. By addressing these challenges and harnessing the unique potential of enzymes, we aspire to pave the way for a brighter future in ND therapeutics. Collaborative efforts between researchers, clinicians, and pharmaceutical companies hold the key to translating these promising scientific breakthroughs into effective therapeutic interventions. The future prospects of enzyme research for NDs are promising, as advancements in biotechnology, nanotechnology, and computational modeling fuel innovation.

References

Agrawal M, Biswas A (2015) Molecular diagnostics of neurodegenerative disorders. Front Mol Biosci 2:54

Baloni P, Funk CC, Readhead B, Price ND (2021) Systems modeling of metabolic dysregulation in neurodegenerative diseases. Curr Opin Pharmacol 60:59–65

Barkovits K, Linden T, Pfeiffer M et al (2020) Blood contamination in CSF and its impact on quantitative analysis of alpha-synuclein. Cells 9(2):370

Bar-Or A, Rieckmann P, Traboulsee A, Yong VW (2011) Targeting progressive neuroaxonal injury: lessons from multiple sclerosis. CNS Drugs 25(9):783–799

Barthel H, Villemagne VL, Drzezga A (2022) Future directions in molecular imaging of neurodegenerative disorders. J Nucl Med 63(1):68–74

Basha FH, Waseem M, Srinivasan H (2021) Cellular and molecular mechanism in neurodegeneration: possible role of neuroprotectants. Cell Biochem Funct 39:613–622. https://doi.org/10.1002/cbf.3630

Bertram L, Tanzi RE (2005) The genetic epidemiology of neurodegenerative disease. J Clin Invest 115(6):1449–1457

Bhat AH, Dar KB, Anees S, Zargar MA, Masood A, Sofi MA, Ganie SA (2015) Oxidative stress, mitochondrial dysfunction and neurodegenerative diseases; a mechanistic insight. Biomed Pharmacother 74:101–110

Bittner B, Richter W, Schmidt J (2018) Subcutaneous administration of biotherapeutics: an overview of current challenges and opportunities. BioDrugs 32:425–440

Bolognesi ML, Cavalli A, Melchiorre C (2009) Memoquin: a multi-target-directed ligand as an innovative therapeutic opportunity for Alzheimer's disease. Neurotherapeutics 6(1):152–162

Brettschneider J, Tredici KD, Lee VM, Trojanowski JQ (2015) Spreading of pathology in neurodegenerative diseases: a focus on human studies. Nat Rev Neurosci 16(2):109–120

Calabrese G, Molzahn C, Mayor T (2022) Protein interaction networks in neurodegenerative diseases: from physiological function to aggregation. J Biol Chem 298(7):102062

Carvey PM, Hendey B, Monahan AJ (2009) The blood-brain barrier in neurodegenerative disease: a rhetorical perspective. J Neurochem 111(2):291–314. https://doi.org/10.1111/j.1471-4159.2009.06319.x

Chen W, Hu Y, Ju D (2020) Gene therapy for neurodegenerative disorders: advances, insights and prospects. Acta Pharm Sin B 10(8):1347–1359

Chien Lab (2024). Available from: https://sites.biochem.umass.edu/chienlab/research/

Cortez L, Sim V (2014) The therapeutic potential of chemical chaperones in protein folding diseases. Prion 8(2):197–202. https://doi.org/10.4161/pri.28938

Cova I, Priori A (2018) Diagnostic biomarkers for Parkinson's disease at a glance: where are we? J Neural Transm 125:1417–1432

Cuny GD (2012) Neurodegenerative diseases: challenges and opportunities. Future Med Chem 4(13):1647–1649

Dokholyan NV, Mohs RC, Bateman RJ (2022) Challenges and progress in research, diagnostics, and therapeutics in Alzheimer's disease and related dementias. Alzheimers Dement 8(1):e12330. https://doi.org/10.1002/trc2.12330

Doroszkiewicz J, Groblewska M, Mroczko B (2022) Molecular biomarkers and their implications for the early diagnosis of selected neurodegenerative diseases. Int J Mol Sci 23(9):4610

Dugger BN, Dickson DW (2017) Pathology of neurodegenerative diseases. Cold Spring Harb Perspect Biol 9(7):a028035

Dunbar CE, High KA, Joung JK, Kohn DB, Ozawa K, Sadelain M (2018) Gene therapy comes of age. Science 359(6372):4672

Ebrahimi Z, Talaei S, Aghamiri S, Goradel NH et al (2020) Overcoming the blood-brain barrier in neurodegenerative disorders and brain tumours. IET Nanobiotechnol 14(6):441–448

Gaetani L, Paoletti FP, Bellomo G et al (2020) CSF and blood biomarkers in neuroinflammatory and neurodegenerative diseases: implications for treatment. Trends Pharmacol Sci 41(12):1023–1037

Gandhi J, Antonelli AC, Afridi A, Vatsia S, Joshi G, Romanov V et al (2019) Protein misfolding and aggregation in neurodegenerative diseases: a review of pathogeneses, novel detection strategies, and potential therapeutics. Rev Neurosci 30(4):339–358

Garibotto V, Albert NL, Barthel H, van Berckel B et al (2021) The approval of a disease-modifying treatment for Alzheimer's disease: impact and consequences for the nuclear medicine community. Eur J Nucl Med Mol Imaging 48:3033–3036

Hansson O (2021) Biomarkers for neurodegenerative diseases. Nat Med 27:954–963

He R, Yan X, Guo J, Xu Q, Tang B, Sun Q (2018) Recent advances in biomarkers for Parkinson's disease. Front Aging Neurosci 10:305

Hu Y, Gao XJ, Zhu Y, Muhammad F, Tan SH, Cao W et al (2018) Nitrogen-doped carbon nanomaterials as highly active and specific peroxidase mimics. Chem Mater 30(18):6431–6439

Ibrahim MM, Gabr MT (2019) Multitarget therapeutic strategies for Alzheimer's disease. Neural Regen Res 14(3):437–440. https://doi.org/10.4103/1673-5374.245463

Jack CR, Bennett DA, Blennow K, Carrillo MC et al (2018) NIA-AA research framework: toward a biological definition of Alzheimer's disease. Alzheimers Dement 14(4):535–562

Kadry H, Noorani B, Cucullo L (2020) A blood–brain barrier overview on structure, function, impairment, and biomarkers of integrity. Fluids Barriers CNS 17(1):69

Kaiafas GC, Papagiannopoulou D, Miliotou AN, Tsingotjidou AS et al (2020) In vivo biodistribution study of TAT-L-Sco2 fusion protein, developed as protein therapeutic for mitochondrial disorders attributed to SCO_2 mutations. Mol Genet Metabol Rep 25:100683

Koníčková D, Menšíková K, Tučková L, Hényková E, Strnad M, Friedecký D et al (2022) Biomarkers of neurodegenerative diseases: biology, taxonomy, clinical relevance, and current research status. Biomedicine 10(7):1760

Kory P, Meduri GU, Varon J, Iglesias J, Marik PE (2021) Review of the emerging evidence demonstrating the efficacy of ivermectin in the prophylaxis and treatment of COVID-19. Am J Ther 28(3):e299–e318

Kumar D, Ashraf G, Bilgrami AL, Imtaiyaz HM (2022) Emerging therapeutic developments in neurodegenerative diseases: a clinical investigation. Drug Discov Today 27(10):103305

Luebke M, Parulekar M, Thomas FP (2023) Fluid biomarkers for the diagnosis of neurodegenerative diseases. Biomark Neuropsychiatry 8:100062

Manoutcharian K, Perez-Garmendia R, Gevorkian G (2017) Recombinant antibody fragments for neurodegenerative diseases. Curr Neuropharmacol 15(5):779–788

Maramai S, Benchekroun M, Gabr MT, Yahiaoui S (2020) Multitarget therapeutic strategies for Alzheimer's disease: review on emerging target combinations. Biomed Res Int 2020:5120230

MDPI (2024) Peripheral biomarkers in neurodegenerative diseases 3.0. Int J Mol Sci. https://www.mdpi.com/journal/ijms/special_issues/biomarkers_ND_3

Medford AJ, Vojvodic A, Hummelshøj JS, Voss J, Abild-Pedersen F, Studt F et al (2015) From the Sabatier principle to a predictive theory of transition-metal heterogeneous catalysis. J Catal 328:36–42

Miliotou AN, Pappas IS, Spyroulias G, Vlachaki E et al (2021) Development of a novel PTD-mediated IVT-mRNA delivery platform for potential protein replacement therapy of metabolic/genetic disorders. Mol Ther Nucl Acids 26:694–710

Miyanohara A, Kamizato K, Juhas S, Juhasova J et al (2016) Potent spinal parenchymal AAV9-mediated gene delivery by subpial injection in adult rats and pigs. Mol Ther 3:16046

Moos WH, Faller DV, Glavas IP, Harpp DN, Kamperi N, Kanara I et al (2021) Pathogenic mitochondrial dysfunction and metabolic abnormalities. Biochem Pharmacol 193:114809

Morales-Navarro S, Prent-Peñaloza L, Rodríguez Núñez, Yeray A, Sánchez-Aros L, Forero-Doria O, González W et al (2019) Theoretical and experimental approaches aimed at drug design targeting neurodegenerative diseases. Handlenet. http://hdl.handle.net/10261/197178

Mukherjee S, Mishra AK, Peer GDG, Bagabir SA et al (2021) The interplay of the unfolded protein response in neurodegenerative diseases: a therapeutic role of curcumin. Front Aging Neurosci 13:767493

NanoTemper Technologies (2020) 3 biggest challenges in neurodegenerative disease research and why you should focus on them. https://nanotempertech.com/blog/3-biggest-challenges-in-neurodegenerative-disease-research-and-why-you-should-focus-on-them/

Noble W, Burns MP (2010) Challenges in neurodegeneration research. Front Psych 1:7

Palaninathan V, Balasubramanian S, Sakthi KD (2015) Molecular imaging and neurodegenerative disorders. Oxford University Press, Oxford, pp 123–134

Parton RF, Vankelecom IF, Casselman MJ, Bezoukhanova CP, Uytterhoeven JB, Jacobs PA (1994) An efficient mimic of cytochrome P-450 from a zeolite-encaged iron complex in a polymer membrane. Nature 370(6490):541–544

Reynaud E (2010) Protein misfolding and degenerative diseases. Nat Educ 3(9):28

Risacher SL, Saykin AJ (2021) Neuroimaging advances in neurologic and neurodegenerative diseases. Neurotherapeutics 18(2):659–660

Ross CA, Poirier MA (2021) Chaperones in neurodegenerative diseases: emerging roles and therapeutic potential. Nat Rev Neurol 17(10):601–613

Siafaka PI, Özcan Bülbül E, Dilsiz P, Karantas ID, Okur ME, Üstündağ ON (2021) Detecting and targeting neurodegenerative disorders using electrospun nanofibrous matrices: current status and applications. J Drug Target 29(5):476–490

Siafaka PI, Okur ME, Erim PD, Çağlar EŞ et al (2022) Protein and gene delivery systems for neurodegenerative disorders: where do we stand today? Pharmaceutics 14(11):2425

Soto C, Pritzkow S (2018) Protein misfolding, aggregation, and conformational strains in neurodegenerative diseases. Nat Neurosci 21:1332–1340

Sweeney P, Park H, Baumann M et al (2017) Protein misfolding in neurodegenerative diseases: implications and strategies. Transl Neurodegener 6:6

Swift IJ, Sogorb-Esteve A, Heller C et al (2021) Fluid biomarkers in frontotemporal dementia: past, present and future. J Neurol Neurosurg Psychiatry 92:204–215

Tikhonova MA, Chang H, Singh S (2023) Editorial: Experimental and innovative approaches to multi-target treatment of Parkinson's and Alzheimer's diseases - volume II. Front Neurosci 17:1171866

Turner AJ (2004) Targeting amyloid-degrading enzymes as therapeutic strategies in neurodegeneration. Ann N Y Acad Sci 1035(1):1–20

Waheed S, Sakariyau R, Obunadike C, Ogunsanmi A (2022) Mechanism of action of chaperones in protein function. Nat Sci 44:333–354

Wang X, Hu Y, Wei H (2016) Nanozymes in bionanotechnology: from sensing to therapeutics and beyond. Inorg Chem Front 3(1):41–60

Wang P, Wang T, Hong J, Yan X, Liang M (2020) Nanozymes: a new disease imaging strategy. Front Bioeng Biotechnol 8:15

Weber WA, Czernin J, Anderson CJ, Badawi RD et al (2020) The future of nuclear medicine, molecular imaging, and theranostics. J Nucl Med 61(2S):263S–272S

Wu J, Li S, Wei H (2018) Multifunctional nanozymes: enzyme-like catalytic activity combined with magnetism and surface plasmon resonance. Nanoscale Horizons 3(4):367–382

Wu J, Wang X, Wang Q, Lou Z, Li S, Zhu Y et al (2019) Nanomaterials with enzyme-like characteristics (nanozymes): next-generation artificial enzymes (II). Chem Soc Rev 48(4):1004–1076

Wu H, Zhou Y, Wang Y, Tong L et al (2021) Current state and future directions of intranasal delivery route for central nervous system disorders: a scientometric and visualization analysis. Front Pharmacol 12:717192

Wu D, Chen Q, Chen X, Han F, Chen Z, Wang Y (2023) The blood–brain barrier: structure, regulation, and drug delivery. Signal Transduct Target Ther 8(1):1–27

Xu R, Zhang S, Wang P, Zhang R, Lin P, Wang Y et al (2024) Nanozyme-based strategies for efficient theranostics of brain diseases. Coord Chem Rev 501:215519–215519
Zetterberg H, Burnham SC (2019) Blood-based molecular biomarkers for Alzheimer's disease. Mol Brain 12(1):1–7
Zhang W, Xiao D, Mao Q et al (2023) Role of neuroinflammation in neurodegeneration development. Sig Transduct Target Ther 8:267
Zhao L, Ren TH, Wang DD (2012) Clinical pharmacology considerations in biologics development. Acta Pharmacol Sin 33(11):1339–1347
Zlokovic BV (2008) The blood-brain barrier in health and chronic neurodegenerative disorders. Neuron 57(2):178–201

The manufacturer's authorised representative in the EU is Springer Nature Customer Service Centre GmbH, Europaplatz 3, 69115 Heidelberg, Germany. If you have any concerns regarding our products, please contact ProductSafety@springernature.com

Printed and bound by CPI Group (UK) Ltd, Croydon, CR0 4YY
01/10/2026
02225910-0001